An Evolutionary Perspective on the Relationship Between Humans and Their Surroundings

Geoengineering, the Purpose of Life & the Nature of the Universe

Neil Paul Cummins

Cranmore Publications

*A catalogue record for this book is available
from the British Library*

ISBN: 978-1-907962-53-0

Published by Cranmore Publications

Reading, England

In this book I seek to give a deeper elucidation of certain aspects of the view which I initially outlined in *Is the Human Species Special?: Why human-induced global warming could be in the interests of life*. There are numerous aspects and dimensions to this view. I hope that the material in this book will convince some people of the merit of the view; however, there is still much more to be suld.

Some of the material in this book has been published elsewhere; for example, the material in Chapter One is a modified version of one of the appendices in *Is the Human Species Special?* By updating this existing material and bringing it together with new material I am able to present many diverse, but interconnected, aspects of the relationship between humans and their evolving surroundings.

Contents

Preface

We live in an epoch in which a violent clash exists. Humans typically believe that they are radically different from their surroundings. Yet, human knowledge has advanced to the point that has enabled the realisation that humans have evolved from their surroundings through a very gradual process of evolution; a process which has been ongoing since the Big Bang. The purpose of this book is to explore the nature of this violent clash.

There are three interrelated aspects to this exploration. Firstly, I will be considering why contemporary humans typically consider themselves to be radically different from their surroundings; this entails a consideration of how the human perceptual apparatus works and of how conceptions of their surroundings are formed within humans. Secondly, I will be considering the likelihood that humans are actually very similar to their surroundings; this entails an exploration of various phenomena such as mind, consciousness, naturalness, awareness, the senses, perception and 'what-it-is-likeness'. Thirdly, I will be considering whether the human species has a special place in the evolutionary process.

Introduction

What is the relationship between humans and their surroundings? The purpose of this book is to explore this question from an evolutionary perspective. I will be claiming that we live in an epoch in which a violent clash exists. In our epoch the *interactions* between a human and their surroundings typically leads to the belief that humans are radically different from their surroundings. However, *human knowledge* has advanced to the point where humans have realised that they have evolved from their surroundings through a gradual process of evolution. The belief that humans are radically different from their surroundings violently clashes with the belief that the entire universe is a slowly evolving entity which very gradually brings forth new arrangements.

There are three interrelated aspects to my exploration. Firstly, I seek to understand why the violent clash exists: Why do contemporary humans typically consider themselves to be radically different from their surroundings? In order to attempt to understand why this is so I consider how the human perceptual apparatus works, the constrained nature of human perceptions, and how conceptions of their surroundings are formed within humans. Secondly, I consider the likelihood that humans could actually be very similar to their surroundings; this entails an exploration of various phenomena such as mind, consciousness, naturalness, awareness, the senses, perception and 'what-it-is-likeness'. Thirdly, I consider whether the human species has a special place in the evolutionary process.

Humans and their surroundings constitute the universe; the universe as a whole, and all of its parts, are in a continuous state of evolution – planets evolve, life evolves, the human species evolves, human culture evolves and individual humans evolve.[1] The human philosophical endeavour is itself a stage in the evolution of the universe. Evolutionary processes have forged the human species – it was evolutionary processes within the non-human universe that brought the human species into existence. Evolutionary interactions between humans and their surroundings also forged the human perceptual apparatus; the structure of this evolved apparatus provides humans with a particular type of perceptual access to their surroundings. Evolutionary processes also cause humans to conceive of their surroundings to be a particular way. So, when humans philosophise about their surroundings they are philosophising about *that* which has endowed them with a particular range of perceptions and conceptions *of itself.*

Of course, when one starts reflecting on the relationship between humans and their surroundings it is possible to end up believing that there are no surroundings; on this extreme view there is no relationship in need of analysis; all that exists is

---

[1] The term 'universe' includes both humans and 'human surroundings'; the term 'human surroundings' refers to the entire universe which surrounds humans but not to humans themselves. I will sometimes be considering the relationship not between all humans and 'human surroundings', but the relationship between an individual human and all of the surroundings which surround this human (which includes all other humans). I will also sometimes use a phrase like: "humans evolved from *their* surroundings"; I find this phrase useful because it makes clear the issues which are being discussed; however, it is, of course, not strictly linguistically correct to say that humans evolved from *their* surroundings, as before humans evolved there were no human surroundings.

oneself – what appears to be surroundings is simply a fallacious creation within one. There is no way of proving that the extreme solipsist is wrong in their belief that there is no relationship in need of analysis. However, when we enter the realm of deep philosophical issues there is no such thing as proof; all that exists are reasons for belief. If I am so deluded that in reality there are no surroundings, then philosophising about those surroundings would clearly be pointless. So, my starting assumption has to be that there are many humans and that there are surroundings surrounding these humans.

Humans have been philosophising about the nature of the relationship between themselves and their surroundings for thousands of years.[2] The vast majority of this philosophising has taken place within a 'static view' of the relationship. By this I mean that the nature of the relationship has largely been considered at a particular moment in time. So, a human looks at a stone and thinks: *what is the relationship between me and this stone?* A human looks at a chimpanzee and thinks: *what is the relationship between me and this chimpanzee?* I believe that this static way of looking at the relationship is very inadequate; it all too easily leads to the belief that humans are very different from their surroundings.

The alternative to the 'static view' is the 'evolutionary perspective'; if one takes this perspective then it is much more likely that one will come to a very different conclusion. To take

---

[2] I take this 'philosophising' to be something which has existed in the vast majority of humans that have come into existence. Some humans spend a very large proportion of their life considering the relationship between humans and their surroundings; other humans simply have a very occasional thought such as: *How similar am I to those birds and to those trees?*

the evolutionary perspective is not just to believe that the human species evolved, and that the universe evolves. To take the evolutionary perspective is to let evolutionary thinking permeate deep down into every aspect of your thought processes. The evolutionary perspective itself can only evolve at a certain time – it requires as a prerequisite a high level of knowledge concerning human surroundings; it is only within the last century that humans have fully come to comprehend the extent of the evolutionary nature of their surroundings. My second assumption is that both humans and all of their surroundings are in a constant state of evolution, and that as part of this evolutionary process humans evolved from their surroundings.

It is my belief that the 'static view' tends to lead to conclusions which create a chasm between humans and their surroundings. To put this slightly differently, given that the existence of chasms in a gradually evolving universe is intellectually problematic, the 'static view' is a creator of philosophical problems. My objective in this book is to cast aside the 'static view' and to use a thoroughly evolutionary perspective to consider many long-standing philosophical issues concerning the relationship between humans and their surroundings.

My hope is that the evolutionary perspective can provide a more accurate description of the relationship between humans and their surroundings than is likely to be arrived at from the perspective of the 'static view'. I would expect that a more accurate description will be more intellectually appealing – will contain fewer mysteries and chasms – than a less accurate description. So, my hope is that the mysteries that currently exist will become less mysterious, and that the hypothesised chasms will significantly shrink. In short, throughout the chapters which follow, I will be exploring the likelihood that humans are far more similar to their surroundings than most humans ordinarily assume themselves to be.

I'll repeat what I said a little earlier: we are in the realm of deep philosophical issues and in this realm we are not expecting to find proof, there will be just reasons for belief. There are almost always differing views on a particular aspect of the relationship between humans and their surroundings. I should reveal my cards at the start and let you know what my starting premise is: *if there are equally good reasons for believing in two opposing positions then I find it more plausible to support the position which entails that humans are more similar to their surroundings.* For example, if there are good reasons to believe that the only part of the universe that has 'attribute x' are humans, and if there are equally good reasons to believe that the entire universe has 'attribute x', then I find it more plausible to believe that the entire universe has 'attribute x'. The reason for this is that I am taking the evolutionary perspective and this perspective entails that the entire universe is a very gradually evolving whole; this is a good reason to expect similarity rather than radical difference.

I will also reveal an interesting conclusion which we will reach. We will see that there are a whole range of areas in which a human can take a view as to the degree of similarity between themselves and their surroundings. I will refer to the opposing views as 'thick' and 'thin'; if a human believes that they are very different from their surroundings they have a 'thick' view, and if they don't they have a 'thin' view. I will be proposing that the 'thick' views are currently dominant and that they have been dominant for most of human history. The widespread tendency to interpret scientific findings so as to support 'thick' views is one contemporary example of this; in the chapters which follow we will see that science is often interpreted as *proving* that the only parts of the universe that contain feelings such as pain, and phenomena such as colours, are brains. Why is this? There is no scientific proof of such claims concerning the brain, and the 'thin' views are intellectually defensible. Our interesting conclusion is that there is an evolutionary reason why this is so. Humans are parts of the evolving universe which have necessarily come to consider themselves to be very different from their surroundings. I will be proposing that this is because the human species evolved to fulfil a purpose which can only be fulfilled if this is so. The evolutionary perspective provides us with a good reason to believe that *in actuality* humans are very similar to their surroundings, and it also provides us with an explanation as to why humans *believe that* they are very different from their surroundings.

In *Chapter One* I will be considering the question of whether humans are wholly natural. I am amazed at how many humans seemingly believe that humans are not natural, or that humans are semi-natural, whereas the entire non-human universe is wholly natural. I am also curious as to why humans raise the question of whether they are natural – the very raising of the question seems to imply that a negative answer was a possibility. In this chapter I make the case that both humans, and all of their surroundings, are wholly natural.

In *Chapter Two* I explore the state of our knowledge concerning the way that our surroundings evolve through time. Many humans seem to believe that humans have a good understanding of biological evolution – they believe that speciation, and all evolution, is underpinned by 'natural selection'; many humans also seem to believe that humans have a good understanding of the non-biological evolutionary processes that are traceable back to the Big Bang. My aim in this chapter is to show that this belief is wholly misplaced; humans cannot fully understand why *any* of their surroundings evolve the way that they do. This conclusion sets the groundwork for my consideration, in *Chapter Eight,* of whether humans have a special place in the evolutionary process.

In *Chapter Three* I consider how the relationship that exists between individual humans and the surroundings which evolved them leads to the creation of particular conceptions of those surroundings (particular 'world views'). I contend that the human perceptual apparatus is inevitably constrained in several ways and that these constraints have the effect of causing humans to believe that they are very different to their surroundings. I explore the relationship between perception and conception, and also the relationship between thought, concepts, perception and the 'surroundings in-themselves'. In this chapter I also outline several different conceptions of human surroundings.

In *Chapter Four* I consider the phenomena of mind and consciousness. The concept of having a mind is one of the key ways that humans seek to carve a division between themselves and the non-human universe, or between themselves and a few other species of animal, and the rest of the universe. In this chapter I address the question of what it means to have a mind. This is important because I conclude that most of the attributes that humans take to be constitutive of having a mind are not constitutive of a mind. This means that only humans and other complex life-forms might have minds, but that all of these other 'mental' (i.e. actually non-mental) attributes could pervade the universe. I also consider the 'mystery of consciousness' and from an evolutionary perspective conclude that it is not so mysterious after all; again, this is because too much is packed into the concept and much of it could quite intelligibly pervade the universe.

In *Chapter Five* I consider the phenomena of 'what-it-is-likeness' and the human senses. I separate the different types of 'what-it-is-likeness' and consider which, if any, are likely to solely exist in humans (or solely exist in humans and some non-human animals) and which are likely to pervade the entire universe. I also consider the human senses and the traditional idea that humans have five senses. It is quite common to think that humans are different from the vast majority of their surroundings because humans can see, hear, taste, touch and smell. I offer a reconceptualisation of the human senses according to which humans only have two senses, whilst the three 'non-senses' pervade the entire universe. I also consider whether the 'what-it-is-likeness' of the two human senses is generated by humans, or whether it pervades human surroundings and exists without being perceived. I conclude that the phenomenon of 'what-it-is-likeness' does not create a division between humans and their surroundings.

In *Chapter Six* I consider the phenomena of awareness, perception and 'what-it-is-likeness'. There are a multitude of possible relationships between these three phenomena. Some of these possible relationships entail that there is a greater chasm between humans and their surroundings than others. I consider these possible relationships and conclude that whilst awareness might be a fairly rare attribute in the universe, there are good reasons to suppose that 'what-it-is-likeness' and perception pervade the entire universe. Furthermore, I contend that 'what-it-is-likeness' and perception are actually a single phenomenon.

This means that a 'what-it-is-likeness'=perceiving (but non-aware) universe might have brought the human species, and its attribute of awareness, into existence.

In *Chapter Seven* I consider the issue of human special-ness: Why have humans come to believe that they are very different from the rest of the life-forms that live on the Earth? Why do humans doubt their naturalness? I also consider whether humans have 'special' unique attributes which their surroundings lack. From the 'static view' I consider a whole range of possible attributes which could create a chasm between humans and their surroundings, and I conclude that there are no 'chasm-creating' attributes.

In *Chapter Eight* I explore the issue of whether humans have a special place in the evolution of their surroundings. It is widely held that not only are humans very different from the surroundings which evolved them, but that they also have no significant relationship with their surroundings. This belief is typically underpinned by the belief that we have sufficient knowledge of evolution to know that humans evolved as some kind of 'fluke' via the process of 'natural selection'. Having undermined this underpinning belief in *Chapter Two,* in this chapter I outline the positive case as to why one should believe that the human species has a relationship of immense impor-tance with the surroundings that evolved it.

Finally, in *Chapter Nine,* I draw some conclusions concern-ing the likely nature of the relationship between humans and their surroundings from an evolutionary perspective.

Chapter 1

Are humans wholly natural?[3]

The definition of natural is 'present in or produced by nature'. Is it not obvious to anyone who thinks about the question of whether humans are wholly natural that humans have been *produced by* nature, and that every fibre of their being and existence is *present in* nature? Surely, a more appropriate question would be: "How can humans possibly doubt that they are wholly natural?"

In a trivial sense, humans – as creators of words – can create a word such as 'natural' and define it in such a way that it excludes humans. The word 'natural' can be opposed to *either* 'artificial' *or* 'supernatural'. However, interestingly, the word 'artificial' is defined as 'made by humans; produced rather than natural'. This definition does not refer to humans *themselves;* rather the word 'artificial' is itself an arbitrary construct, a word of use in human communication because it enables the *productive activities of humans* to be referred to. There is no notion in the word 'artificial' that humans themselves are not natural. Furthermore, there is no implication that in the universe itself there is a fundamental division that the word 'artificial' refers to.

---

[3] This chapter is a slightly modified version of an essay which won the 2008 international prize essay competition (€1000) run by the Spinoza-Gesellschaft. The essay was presented to the international conference of the Society in Marburg, Germany, in September 2008.

It is simply of use to humans to have a word that labels the results of their productive activities.

The notion of 'human production' is actually a deeply problematic one. If one gives the matter no real thought then the distinction between the 'natural' and the 'artificial' seems to be obvious, but reflection reveals otherwise. It is obviously the case that before the human species evolved nothing was artificial. It also seems obvious that objects such as the Sun and the planet Jupiter are not artificial. But when we focus on the Earth, then it is hard to identify anything that is truly natural. Let us consider a tree that is growing in a rainforest, and a tree that is produced in a human factory. One would be tempted to call the former 'natural' and the latter a 'produced artefact'. However, when one learns that the rainforest tree has the particular attributes that it has because humans soaked the surrounding ground with nutrients and breathed additional carbon dioxide into the vicinity of the tree, then one might have to concede that the tree is an artefact.

Similarly, a human-constructed wigwam-shaped structure composed of branches would be a 'produced artefact', but a single branch lying on the ground under a tree would be considered 'natural'. Even if a human steps on the branch and breaks it the branch would still be considered 'natural'. However, there is no difference in *the universe itself* between branches which have been modified by being stepped on, and branches which have been moved into a wigwam structure. At a larger scale human activities have modified the climate and atmosphere of the entire planet thereby making the concept of the 'natural', as opposed to the 'artificial', largely redundant when it comes to the biosphere of the Earth. So, there is no meaningful distinction *in the universe itself* between the 'artificial' and the 'natural'.

The word 'supernatural' is defined as: 'of or relating to existence outside the natural world'. It has to be questionable whether this word has any meaning whatsoever – surely all that exists is the natural world – *nothing* exists outside it. The word 'supernatural' is also used to refer to 'a power that seems to violate or go beyond natural forces', and also 'of or relating to the miraculous'. These descriptions are instructive because they imply that humans use the word 'supernatural' to refer to those parts of the universe around them that they cannot comprehend. So, in the past a total eclipse of the sun would have been referred to as a supernatural event, a miracle. But now such an event is simply considered to be a natural occurrence. This means that the word 'supernatural' can be thought of as delineating human *understanding* of how the universe works from how it *actually* works; it doesn't imply that in reality parts of the universe are not natural. A complete description of the universe would include solely 'natural' phenomena; there wouldn't be any 'supernatural' phenomena.

So, the word 'artificial' simply refers to the productive activities of humans, and the word 'supernatural' can be thought of as delineating the limits of human understanding. These terms are perfectly compatible with the belief that humans are wholly natural, and the belief that all that exists is the natural world. Indeed, if one denies this and asserts that the universe, and the Earth, contains natural parts and non-natural parts then one faces a seemingly inextricable problem. One has to try to untangle the complex and intricate way in which the activities of humans have modified their surroundings in order to separate what is natural from what is not natural. If one attempted to do this then one would eventually conclude that the entire biosphere of the Earth is *not* natural. It is surely more sensible to conclude that the whole of the Earth is natural, and that *the word 'artificial' refers to things that, whilst produced by humans, are still natural.* Alternatively, if one really believes that *the universe itself* is split into the natural and the super-natural, then one has to specify exactly what these supernatural entities are. History suggests that there are no such entities because as understanding increases the 'supernatural' gets reclassified as natural.

Nevertheless, the very fact that humans can doubt whether they are natural is clearly of interest. To understand the existence of this doubt we need to deconstruct the word 'natural'. Natural has been defined as: 'present in or produced by nature'. The word 'nature' itself has not yet been defined, but is contemporarily defined as: 'the material world and its phenomena'. This definition of nature sheds light on why humans could possibly doubt that they are completely natural. It is the contemporary *conception* of nature as the 'material world' that causes humans to doubt whether they are 'natural' because humans themselves don't seem to be solely composed of 'matter'.

In the face of this doubt there are two 'naturalistic' options. Firstly, one could hold that the term 'matter' is an accurate description of most human surroundings, but that humans are not solely made of 'matter', and therefore that 'matter' has the ability to *evolve* entities such as humans which are not solely made of 'matter'. Secondly, one could hold that the term 'matter' is vacuous – there are no parts of the universe that fit the term. In other words, one can hold that human knowledge of 'nature' is still at a primitive level and that the label 'material world' is deeply misleading; it could be the case that all of nature has states analogous to those in humans. From the evolutionary perspective – according to which the universe slowly changes over time, gradually bringing forth new arrangements – the second option is clearly the most appealing. Both of the options are 'naturalistic' options; both entail that humans *are* wholly natural. However, if one has the former belief then it is much more likely that one will consider oneself to *not* be natural.

Most of the above analysis relies on the definition of words – definitions that can help explain why humans might not consider themselves to be natural, *given* the definition of natural. However, there is a much deeper issue as the definitions themselves are clearly expressions of a pre-existing conceptual framework. The real issue is why the human conceptual framework in the current epoch is such that nature is defined as the 'material world and its phenomena'. Why do most contemporary humans consider themselves to be fundamentally opposed to their surroundings? Could it be that it is a fundamental characteristic of what it is to be human to conceive of oneself as opposed to one's surroundings? The answer to this question depends on what it means *to be human*. Is *to be human* to be part of a particular biological species? If so, then one could be part of that species and conceive of oneself as not opposed to one's surroundings. Perhaps there is a more subtle answer to the question. Perhaps *to be human* is something that transcends biological classification. Perhaps *to be human* is to be a part of nature that considers itself to be not natural. If this is so, then when a particular biological entity loses its sense of opposition – of separateness – of non-naturalness – then it will cease to be human.

There are clearly two distinct issues. Firstly, given that the word natural means 'present in or produced by the material world and its phenomena', what could it possibly mean for humans to be 'not natural'? Secondly, if humans are wholly natural why do they doubt their 'naturalness'? Why do they consider themselves to be opposed to their surroundings? These two questions are obviously closely interrelated because it is the belief in an opposition between humans and their surroundings which leads to the conceptualizing of those surroundings as 'matter'. If the belief in an opposition didn't exist then the natural world itself would, no doubt, not be conceptualized as 'matter'. Rather, it would be conceptualized in such a way that the attributes of humans and their surroundings are tightly coupled.

1.1 Human conceptions of the natural world

It goes without saying that human conceptions of the natural world have varied immensely through time, and that they also vary between different cultures in the contemporary epoch. It is the dominant contemporary 'western view' of the natural world which I will focus on here. This view establishes a particularly sharp division between humans and their surroundings, a division that is so deep that the naturalness of humans can be seriously doubted. According to the 'western view' the natural world is a mechanism the activities of which have been increasingly accurately predicted through science. The operations of the non-living world, and much of the living world, are conceived of as predictable and as wholly devoid of qualitative feeling, mind, intentionality and consciousness. In contrast, humans have 'free will' to act in ways which cannot be predicted by science, and have qualitative feeling, mind, consciousness and intentionality.

Of course, there are those who live in the 'west' who don't subscribe to the 'western view'. Some humans hold that the entire natural world has intentionality, or qualitative feeling, or consciousness, or that quantum physics shows that the entire natural world has freedom. Others argue that humans themselves don't have free will – it is simply an illusion. What is one to make of the claims of these humans who oppose the 'western view'? The issues raised are very deep and many seem to be unanswerable. Who could say whether states of about-ness/intentionality exist when atoms interact to form molecules? Who could say whether these interactions themselves entail qualitative feeling? Who could say whether there is any kind of consciousness present in these interactions? Who could say whether these interactions are manifestations of freedom? And who could possibly know if every thought that they have ever had is determined, and in theory predictable before they had it?

I take it that no-one has adequate answers to these questions. Therefore, the question needs to be asked as to why the 'western view' sides with the 'oppositions' – the answers to the questions which lead to an opposition between humans and their surroundings. This 'siding' clearly says nothing about the natural world *itself* – all it reveals is the way in which humans conceive of themselves in relation to their surroundings.

I obviously do not claim to have answers to the above questions. However, I do not see any good reason why, given that humans were produced by nature – given that humans evolved from their surroundings – that there should be a great chasm between the attributes of humans and their surroundings. This doesn't mean that humans cannot have unique attributes, just as a human eye has attributes that a human finger lacks. It is possible that, for example, the high level of thought that occurs in humans is a unique attribute of humans. But, just as the body of a human is pervaded by unique attributes, and just as the attributes of mercury are very different to those of helium, uniqueness doesn't entail a fundamental division in the universe. It is surely the case that the vast majority of the attributes of humans are shared by the entire natural world, whilst every phenomenon in the natural world (including humans) also has some kind of uniqueness when analyzed in detail.

It could be argued that the question: "Are humans wholly natural?" should be replaced with the question: "How many of the attributes of humans are not present in the non-human world?" At a first glance this question would seem to provide some kind of an answer to the former question. If it was concluded that humans have a plethora of attributes that are not present in the non-human world then this would seemingly lend support to the idea that humans are largely not natural. Whilst, contrarily, if it was concluded that there are hardly any 'unique' human attributes this would seemingly imply that humans are very largely natural. However, it is clear that the latter question cannot provide any kind of answer to the former question. Whilst the latter question is a valid question to ask it is comparable to asking the question: "How many of the attributes of mercury are not present in the non-mercury world?" There clearly are attributes of mercury that are not present in the non-mercury world, because it is the presence of these attributes that makes mercury mercury. But it would be nonsensical to conclude from this that the question: "Is mercury wholly natural?" is a sensible question to ask. It is simply the case that within the natural world there are differences in the attributes of the various phenomena that exist; some phenomena will closely resemble others, and some will not.

It is time to consider the place of non-human animals. We have seen that 'natural' is defined as 'present in or produced by the material world and its phenomena', and is conceptually opposed to both 'artificial' and 'supernatural'. Given these 'oppositions' what are we to make of the 'naturalness' of other animals? We know that all living things modify their surrounding environment, and that many species of animals are very human-like in their activities. For example, chimpanzees are tool-users, beavers construct dams, and birds construct nests. These activities and modifications of their surroundings by non-human animals are clearly not 'artificial', because artificial is defined as 'made by humans; produced rather than natural'. They are surely also not 'supernatural', which is defined as: 'of or relating to existence outside the natural world'. These activities are surely wholly natural.

I take it that it would be indefensible to describe non-human animals themselves, or their constructions, as anything other than wholly 'natural'. But I also take it that it would be woefully inadequate to describe animals such as chimpanzees simply as 'matter'; there are close links between the attributes of humans and the attributes of many non-human animals. But, it has been accepted that chimpanzees, beavers and birds, and their activities, are wholly 'natural'. This means that they are wholly produced by or present in the material world; they are either 'matter' or the result of the interactions of 'matter'. So, as with humans, there is a clear tension here – chimpanzees and beavers are both 'natural' *and* 'more than matter' at the same time.

This tension gets to the heart of the issue of the relation between humans and the 'natural world'. In fact, other animals are an 'intermediary' between humans and phenomena such as mercury. It seems easy to assert that mercury is both wholly natural and 'matter'. But when it comes to an animal such as a chimpanzee, whilst it is easy to assert that it is wholly natural, it also seems to be correct to assert that it is 'more than matter'. This tension gets continued into the realm of humans, because humans are also surely 'more than matter'. It is this tension which leads to the conclusion that maybe humans are not wholly 'natural' because the 'natural' is fundamentally 'matter'. But if we accept this conclusion then we surely also have to accept that some non-human animals are also not wholly 'natural'. This is surely unacceptable.

What is the alternative? If the similarities between humans and certain species of non-human animals are accepted, which they surely should be, then it has to be accepted that if these non-human animals and their activities are wholly natural, then so are humans and their activities. Furthermore, we should change our conceptualization and definition of 'natural' by ridding it of the notion of the 'material' world. In other words, we should initially accept our ignorance about the fundamental nature of reality, and then we should conclude that the word 'material' is vacuous. This means that we can then define 'natural' as: 'present in or produced by the world and its phenomena'. This definition quite helpfully rids us of the notion of the 'supernatural'. It also leaves open the possibility that there is a tight coupling between the attributes of humans and their surroundings. If this conceptualization became the dominant view of the human-nature relationship, rather than the 'western view', then surely humans would consider themselves to be wholly natural. In the present epoch humans doubt their 'naturalness'.

1.2 Why do humans doubt their 'naturalness'?

It is slightly paradoxical that humans can on the one hand talk of the evolution of all species from a common ancestor, and the evolution of the entire universe from the Big Bang, and yet, on the other hand, they can doubt their 'naturalness'. Perhaps this is so because humans are that part of the universe which *of necessity* considers itself to be not natural. In other words, in the form of humans, the universe has produced a kind of 'reflective mirror' which enables it to do things that are impossible without such a mirror. A useful analogy is the hairdresser, who is capable of creating the perfect haircut for their clients without a mirror, but who can only produce a dreadful mess on their own hair without the aid of a mirror. If the hairdresser produces a mirror they will gain the ability to perfectly cut their own hair, just as the universe clearly gains abilities through producing humans.

If this is right – if humans have unique abilities due to not considering themselves to be natural – then this means that humans will doubt their naturalness. To be human *is* to consider oneself to not be natural; to be opposed to one's surroundings; to be alienated from the rest of the universe. According to this view it is inevitable that humans will doubt their 'naturalness'. It is when this doubt arose that the 'human' was born. What exactly does it mean to be a human? What exactly does it mean for a part of the universe to doubt its naturalness? What it means is that one will consider humans to be very different to the entire non-human universe; one will conceptualise the entire non-human universe as 'nature' and then will ask the question: Are humans natural? One will consider humans as superior to the entire non-human universe – *these other life-forms on the Earth, they are non-human, they are just animals.* In *Chapter Seven* we will further consider why humans ask this question (Are humans natural?); I will claim that humans ask this question because they evolve in surroundings which are pervaded with advanced tools, and these tools give rise to a 'sense of specialness'. Then, in *Chapter Eight,* I will explain exactly which abilities the universe has gained by bringing humans into existence.

What of the future? Could all of the parts of the universe that we call 'biological humans' stop doubting their naturalness? If this scenario were to transpire then it would be open to debate whether or not there were any 'humans'. I have suggested that what it is *to be human* could be something that transcends biological classification; the elimination of doubt could simultaneously be the elimination of the human. The elimination of the human in the future is a possibility.[4] In this chapter, rather than to speculate too much about the future, my aim has been to focus on the present. In the present humans exist, and it seems that in being that part of the universe which has evolved to doubt its naturalness that humans have a valuable place in the universe.

1.3 Concluding remarks

I have claimed that humans are completely and utterly natural in every fibre of their being. The word 'artificial' is simply a label that is used in human communication to refer to the productive activities of humans. The word 'supernatural' simply delineates humans understanding of their surroundings from the way those surroundings actually are.

I have claimed that some species of non-human animals are sufficiently similar to humans that if humans aren't wholly natural then this implies that these non-human species are also not wholly natural. I have also claimed that it would be futile to attempt to divide the phenomena of the Earth up into the 'natural' and the 'artificial'; the inextricability of the 'artifi-

---

[4] We will consider this possibility further in *Chapter Eight* where we explore the view that when the universe evolved humans it created a disharmony, and that this will be followed by a regained harmony when the human is 'eliminated'.

cial'/'natural' division forces one to accept that everything is natural.

I have proposed that a more perplexing question than: "Are humans wholly natural?" is, "How can humans possibly doubt that they are completely natural?" I have suggested that humans in the contemporary epoch consider themselves to be not natural because they consider the overwhelming majority of their surroundings to be 'matter'. From the evolutionary perspective I have claimed that there are good reasons to suppose that *in reality* the attributes of humans and their surroundings are actually tightly coupled. Furthermore, I have suggested that humans are that part of the evolving universe which comes to consider itself to be not natural. It is this belief, this doubt, which gives rise to the human, and it gives humans their unique position in the evolutionary process. Humans and doubt are inextricably linked. Nevertheless, humans are wholly and utterly natural.

Chapter 2

Evolution

the interpretations surrounding the brute fact of evolution
remain contentious, controversial, fractious, and
acrimonious.

Simon Conway Morris (2005, p.2)

It is very widely, and increasingly, accepted that evolution occurs. However, to say this, in itself, is not to say very much. Evolution is widely accepted to occur both in the biological realm and at the level of the entire universe. To say that evolution occurs at the level of the entire universe is simply to say that things change – parts of the universe bring forth new things – they take on new forms; this process of novel bringing forths is typically traced back to the event known as the Big Bang. To say that evolution occurs in the biological realm is simply to say that life-forms change; the epitome of change in the biological realm is the bringing forth of a new species. The bringing forths in the biological realm are a subset of the bringing forths in the wider universe. To say that evolution occurs is obviously a very different thing to saying that humans have a good understanding of how evolution occurs. Indeed, one can quite reasonably, and sensibly, assert that one is convinced that evolution occurs but that one is utterly clueless as to the nature of the processes underlying evolution.

In this chapter I have four objectives. The first objective is to consider a number of different possible mechanisms of biological evolution. The second objective is to consider which of these mechanisms, if any, plays the key role in the creation of new species. The third objective is to consider evolution in the non-biological realm. The fourth objective is to consider the limits of human knowledge concerning both biological and non-biological evolution.

The theory of biological evolution by common descent is now generally accepted as a brute fact. It was Charles Darwin who established beyond any reasonable doubt that evolution occurs. However, there are three meanings of evolution: *evolution as fact* – species are not fixed but arise out of and develop into other species, *evolution as path* – the actual routes that evolution has taken, and *evolution as mechanism* – the power that lies behind evolutionary change. Darwin himself had very little to say about the path of evolution, whilst the potency of his postulated evolutionary mechanism – 'natural selection' – has been seriously questioned.

So, I am taking it for granted that evolution is a fact, and my objective is to critique various interpretations of 'evolution as path' and 'evolution as mechanism'. The current dominant paradigm of evolution is the 'modern synthesis': a synthesis of Darwinian natural selection with Mendelian genetic inheritance. This paradigm stresses that the *path* of evolution is gradualistic: very small cumulative changes in organisms give rise to adaptation. It also postulates that the *mechanism* of evolution is solely the natural selection of random gene mutations: all of the variation that gives rise to speciation and adaptation is argued to arise from the random mutation and reproductive shuffling of genes.

Various other positions have been suggested with regards to both the path and the mechanism of evolution. This is not surprising given that the fossil record itself gives no support to a gradualistic path, and given that there have been no recorded cases of speciation through random mutation. I will be considering the arguments of two of the major contemporary alternative positions to see whether they make a convincing case against the 'modern synthesis' assumptions regarding 'evolution as path' and 'evolution as mechanism'. The first of the alternative positions is Developmental Systems Theory which postulates that the object of natural selection is the whole life cycle of an organism. The second is Symbiogenesis Theory which postulates that natural selection has only a minor role in evolution.

In *Section 2.1* I outline the claims of the dominant 'modern synthesis' paradigm, with the concentration being on 'selfish gene neo-Darwinism'. In *Sections 2.2* and *2.3* I outline the alternative positions, and I elucidate their arguments as to why the 'modern synthesis' is flawed in its claims regarding the path and/or the mechanism of evolution. In *Section 2.4* I draw together the various critiques so that conclusions can be made about the validity of the 'modern synthesis' claims regarding 'evolution as path' and 'evolution as mechanism'. In *Section 2.5* I consider evolution in the non-biological realm and conclude that humans cannot fully understand either biological evolution or non-biological evolution. In *Section 2.6* I consider and reject the opposing idea that such understanding is possible. Finally, in *Section 2.7,* I draw some conclusions.

2.1 *The neo-Darwinian modern synthesis*

Charles Darwin proposed that the mechanism of evolution was natural variation and selection. He realized that more organisms are born than can survive and reproduce, and was thus led to hypothesize a struggle for survival. As there is obviously variation between organisms, those with advantageous traits will be those that survive. The heritable nature of the traits of the survivors leads to evolution, adaptation and speciation. However, Darwin "could say little about the nature and causes of hereditary variation" (Jablonka and Lamb, 2005, p.10).

It was the rediscovery of Gregor Mendel's model of genetic inheritance, in the early twentieth century, that provided a plausible model of hereditary variation. Mendel's "first law" asserts that the two alleles of each hereditary unit (gene) separate during gamete formation. His "second law" asserts that allele segregation in differing parts is wholly independent. It follows that there will be an immense amount of heritable variation in the gametes. The realization that this gene variation could be the basis for Darwinian natural selection led to the formulation of the neo-Darwinian 'modern synthesis'.

In the 'modern synthesis' natural selection solely acts on variations in persisting elements. Populations, individual organisms and chromosomes do not persist; what does persist is small amounts of genetic material that are not broken up in meiosis. Richard Dawkins calls these stretches of genetic material the "immortal replicators". This focus on the selection of persisting elements means that there is a fundamental distinction in the 'modern synthesis' between the genotype/replicator on the one hand, and the phenotype/vehicle on the other.

Evolution as mechanism is envisioned as simply a change in the genetic composition of populations. Heredity occurs through the transmission of chromosomal germ-line genes, which carry trait information. Variation results from allele shuffling in meiosis and random mutations, with each allele having a small phenotypic effect. It is the selection of adapted phenotypes which causes the increasing prevalence of their respective genotypes.

Dawkins is a strong advocate of the view that the units of adaptive selection are genes rather than organisms. His theory of the "selfish gene" implies that sometimes the interests of the genes within an organism will be different to the interests of the organism itself. In *The Selfish Gene* Dawkins centers his arguments on the fragmenting effects of meiosis, which he argues is clear evidence that organisms cannot be the replicators that natural selection works on. Natural selection requires persisting elements over many generations, organisms only last for a single generation. Dawkins stresses that genes have only *potential* immortality; a gene can last for millions of years or only a single generation. The reasons lying behind these differential survival rates of genes are central to his arguments (1999a, p.2, p.36):

Like successful Chicago gangsters, our genes have survived, in some cases for millions of years, in a highly competitive world. This entitles us to expect certain qualities in our genes. I shall argue that a predominant quality to be expected in a successful gene is ruthless selfishness. This gene selfishness will usually give rise to selfishness in individual behaviour.

The few new ones that succeed do so partly because they are lucky, but mainly because they have what it takes, and that means they are good at making survival machines. They have an effect on the embryonic development of each successive body in which they find themselves, such that that body is a little bit more likely to live and reproduce than it would have been under the influence of the rival gene or allele.

So competition is between genes which are vying for a chromosomal slot at the expense of their alleles. Dawkins uses what he describes as a "fading out" definition of a gene. The word "gene" refers to a stretch of genetic material that survives in the form of lots of copies for a significant period of evolutionary time. The smaller the stretch of genetic material, the less likely it is to be divided, and the more copies of it there are likely to be. This causes Dawkins (1999a, p.33) to assert that *The Selfish Gene* should be called: "*The slightly selfish big bit of chromosome and the even more selfish little bit of chromosome.*"

This gene's-eye view of allele competition and multiple gene copies has two obvious implications. Firstly, there must be a selfish explanation for apparently altruistic behaviour at the level of the organism: "relatives share a substantial proportion of their genes. Each selfish gene therefore has its loyalties divided between different bodies" (Dawkins, 1999a, p.87). Secondly, there will be Machiavellian *within* organism strategies as genes vie for survival: "There are even genes – called mutators – that manipulate the rates of copying-errors in other genes" (Dawkins, 1999a, p.44). The mutator genes selfishly spread through the gene pool at the expense of those genes which are disadvantaged by the miscopying.

How is heritable variation hypothesised to originate? Dawkins (1999a, p.205) claims that: "A body is the genes' way of preserving the genes unaltered." This means that: "A monkey is a machine which preserves genes up trees, a fish is a machine which preserves genes in the water; there is even a small worm which preserves genes in German beer mats" (Dawkins, 1999a, p.204). This means that evolution is an unwanted accident. The genes do not want to evolve new vehicles; they simply seek to preserve themselves; it is undesirable random mutations which have led to variation, adaptation, evolution, and speciation.

In *The Extended Phenotype* Dawkins expands and enriches the selfish-gene theory. He realizes that his concentration on meiosis in *The Selfish Gene* was missing the point because it implies that an asexually reproducing organism would be a replicator in itself, a giant gene. This means that (Dawkins, 1999a, p.274):

> *Superficially, successive generations of stick-insect bodies appear to constitute a lineage of replicas. But if you change one member of the lineage (for instance by removing a leg), the change is not passed on down the lineage. By contrast, if you experimentally change one member of the lineage of genomes (for instance by X-rays), the change will be passed on down the lineage. This, rather than the fragmenting effects of meiosis, is the fundamental reason for saying that the individual organism is not the 'unit of selection' – not a true replicator.*

So, the focus on genes as the units of selection in the 'modern synthesis' evolutionary mechanism rests on the Central Dogma of molecular biology – that information cannot move from proteins to DNA and RNA. Dawkins has been accused of being a genetic determinist; in response to this he stresses that there is a difference between evolution and development. Evolution is an inflexible process of genes replicating themselves. Development concerns the influence of genes on phenotypes, which is exceedingly flexible. Thus (Dawkins, 1999b, p.38):

A gene 'for' A in environment X may well turn out to be a gene for B in environment Y. It is simply meaningless to speak of an absolute, context-free, phenotypic effect of a given gene.

Therefore, it is clear that Dawkins accepts that there is not a perfect correspondence between genotype and phenotype. However, he also claims that (1999b, p.28):

The statement, 'genes for performing behaviour X are favoured over genes for not performing X' has a vaguely naive and unprofessional ring to it...To say 'individuals that perform X are fitter than individuals that do not perform X' sounds much more respectable...But the two sentences are exactly equivalent in meaning. The second one says nothing that the first does not say more clearly.

So, Dawkins claims that both the phenotypic effects of genes are environment-dependent, and that a phenotypic trait for X is 'exactly equivalent' to a gene for X. It is hard to see how a gene, if its phenotypic effect is environment-dependent, can in any meaningful way be said to be 'exactly equivalent' to a phenotypic trait. We will see that this point is the focus of the Developmental Systems Theorists belief that the 'modern synthesis' is preformationist. Dawkins attempts to make the statements consistent by arguing that we need to focus on *differences* and *relative* phenotypic effects. He argues (1999b, p.195) that (with G = genotype and P = phenotype):

> *there is a statistical tendency for individuals with G_1 to be more likely than individuals with G_2 to show P_1 (rather than P_2). Of course there is no need to demand that P_1 should always be associated with G_1, nor that G_1 should always lead to P_1: in the real world outside logic textbooks, the simple concepts of 'necessary' and 'sufficient' must usually be replaced by statistical equivalents.*

Therefore, Dawkins's claim that there are 'genes for performing behaviour X' is only a statistical tendency; whilst the 'exact equivalence' can only relate to one particular individual organism with a particular environmental history; it is a highly contingent equivalence. So, whilst Dawkins's position is that there is a statistical tendency for alleles likely to produce certain phenotypic effects to increase in the gene pool by natural selection, this is clearly an environmentally contingent tendency.

Dawkins also seeks to defend his gene selectionism by arguing that there are several ways in which the effects of genes can extend outside their vehicles, and thereby increase their statistical prevalence in the gene pool. The first of these extended phenotypic effects is the construction of animal artefacts. These artefacts – such as the construction of caddis-fly houses – are tools through which genes enhance their chances of getting into the next generation. There can also be shared interests in a single artefact, as when a beaver dam is part of the extended phenotype of several beavers.

The second group of effects relate to parasites and their hosts. Dawkins (1999b, p.210) claims that, "it is logically sensible to regard parasite genes as having phenotypic expression in host bodies and behaviour." An example of this is the case of the 'brainworm' *Dicrocelium dendriticium*. The 'brainworm' burrows into the suboesophagael ganglion of an ant's brain and is thereby able to control the ant's behaviour. The result is that, "infected ants climb to the top of grass stems, clamp their jaws in the plant and remain immobile as if asleep" (Dawkins, 1999b, p.218). This is maladaptive behaviour for the ant – the uninfected ants have retreated to their nest – but it enhances the worm's chances of being eaten by its definitive host.

The final group of extended phenotypic effects is 'action at a distance'. An example is the "Bruce Effect" which is an effect that a male mouse can have on a female mouse that has just been inseminated by another male: chemical exposure blocks the pregnancy. This means that: "Abortion in female mice, according to this hypothesis, is a phenotypic effect of a gene in male mice" (Dawkins, 1999b, p.231). A further example is the cuckoo, which "uses a supernormally bright gape to inject its control into the reed warbler's nervous system via the eyes. It uses an especially loud begging cry to control the reed warbler's nervous system via the ears. Cuckoo genes, in exerting their developmental power over host phenotypes, have to rely on action at a distance" (Dawkins, 1999b, p.227).

The central claim of the extended phenotype theory is that: "An animal's behaviour tends to maximize the survival of the genes 'for' that behaviour, whether or not those genes happen to be in the body of the particular animal performing it" (Dawkins, 1999b, p.233). The range of cases that Dawkins refers to gives strong support for this claim. In many cases the behaviour of an organism is maladaptive in terms of its survival chances, whilst being adaptive for the genes that initiate the behaviour. It follows that the unit of selection is perhaps better thought of as the gene rather than the organism.

So, the mechanism of evolution in the neo-Darwinian 'modern synthesis' is the unwelcome mutation of genes, which can be thought of as selfish and as having extended phenotypic effects. A crucial aspect of this mechanism, which relates 'evolution as mechanism' to 'evolution as path', is that evolution proceeds in a gradualistic manner through accumulating small beneficial changes over time. This causes Dawkins (2006, p.43) to argue that:

Each successive change in the gradual evolutionary process was simple enough, relative to its predecessor, to have arisen by chance. But the whole sequence of cumulative steps constitutes anything but a chance process, when you consider the complexity of the final end-product relative to the original starting point. The cumulative process is directed by non-random survival.

Dawkins argues that the alternative is that a single macromutation could turn bare skin into a fully functioning human eye. This is as likely as "a hurricane blowing though a junkyard and chancing to assemble a Boeing 747" (Dawkins, 2006, p.234). This is 'saltationism' and Dawkins has two arguments against it. Firstly, the larger a mutation is the more likely it is to be harmful: "if we consider mutations of ever-increasing magnitude, there will come a point when, the larger the mutation is, the less likely it is to be beneficial; while if we consider mutations of ever-decreasing magnitude, there will come a point when the chance of a mutation's being beneficial is 50 per cent" (Dawkins, 2006, p.233). Secondly, the *bare skin–eye macromutation* entails such a "large number of improvements, their joint occurrence becomes so improbable as to be, to all intents and purposes, impossible" (Dawkins, 2006, p.234).

So, in the 'modern synthesis' adaptive evolutionary change comes from very small favourable random gene mutations which are accumulated over time. Natural selection is a constructive force because (Dawkins, 2006, p.193): "genes will be favoured if they are good at cooperating with other genes in the same gene pool", and because (Dawkins, 2006, p.193): "arms races... [propel] evolution in directions that we recognize as 'progressive', complex 'design'." Contrarily, any large macromutations will be harmful and will thus be eliminated by negative selection. This evolutionary mechanism leads to expectations of 'evolution as path'.

The actual evolutionary paths that occurred in the past are clearly a matter of speculation. The 'modern synthesis' asserts that there was gradualistic change with small accumulated adaptive traits forming in response to geographical isolation; after sufficient time this is hypothesised to lead to speciation. So, there was a common ancestor, and there has been common descent through purely branching phylogenies ever since. This implies that our initial expectations of the fossil record would be to find a smooth and gradual transition between species, assuming of course that the record was complete.

In reality the fossil record reveals trends that are extremely jerky, far from the expected smoothness. It was assumed by Darwin that the actual fossil record is jerky because it is incomplete, and that a complete record would show a smooth transition between species. There is another alternative, because, as Dawkins (2006, p.230) claims: "It is conceivable that there really never were any intermediates; conceivable that large evolutionary changes took place in a single generation." In the discussion on gradualism we saw that Dawkins rejects this possibility because it relies on saltationist macromutations which he claims will be unfavourable.

The theory of punctuated equilibrium – originated by Niles Eldredge and Stephen Jay Gould – has often been cited as evidence against gradualism. In the punctuated equilibrium theory evolution occurs in sudden bursts which are punctuated by long periods of stasis. These 'bursts' have often been associated with 'saltationism'; however, Dawkins argues that they are two very distinct things because 'bursts' in the fossil record could equate to thousands of generations of a creature's existence. Dawkins (2006, p.241) does acknowledge that Eldredge and Gould: "saw analogies between themselves and the old schools of 'catastrophism' and 'saltationism'." However, because Dawkins personally finds this inconceivable he claims that (2006, p.241): "Eldredge and Gould are not saltationists." Despite their own analogies, they *must* have only been talking about differential speeds of evolutionary gradualism.

So, how does the 'modern synthesis' explain the fossil record? Dawkins argues that an incomplete fossil record is exactly what would be expected if gradualistic evolution by natural selection occurred. This is because in the 'modern synthesis' geographical separation "is the main process by which new species come into existence" (Dawkins, 2006, p.239). It follows that the fossils at any one location will not reveal information about an 'evolutionary event'; they will reveal information about a 'migrational event'.

Dawkins asks us to imagine an ancestral species in one location; he then postulates that a few members of the species become geographically isolated. What expectations would we have concerning the fossil record? The isolated members will accumulate random mutations that the non-isolated members do not; after sufficient time the isolated members will become a new species. A change that enables the geographical isolation to be transcended will enable the isolated members (now a new species) to return to their ancestral home; the descendant species outcompetes the ancestral species and replaces it. Clearly, the fossil record at this location would be expected to show a jerky transition from the ancestral species to the descendent species (Dawkins, 2006, pp.237-9).

So, to recapitulate, in the 'modern synthesis' the path of evolution is one of purely branching phylogenies, and these speciation events largely result from geographical separation. Changing conditions and 'migrational events' mean that the fossil record will not be smooth. It seems that either a perfectly smooth or an extremely jerky fossil record provides support for the 'modern synthesis'.

2.2 Symbiogenesis

The theory of symbiogenesis has a radically different view of 'evolution as mechanism' and 'evolution as path' to the 'modern synthesis'. According to this view symbiogenesis is the driving force behind speciation and evolution, and it is also the provider of Darwin's heritable variation; natural selection has a role in evolution, but it is only a minor role.

It is important to understand the difference between symbioses and symbiogenesis. Lynn Margulis and Dorion Sagan (2002, p.12) define symbioses as: "long-term physical associations. Different types of organisms stick together and fuse to make a third kind of organism". These associations are very common and are acknowledged by the 'modern synthesis'. Symbiogenesis is something that can result from a symbiosis. Margulis and Sagan (2002, p.12) explain symbiogenesis as follows:

> *As members of two species respond over time to each other's presence, exploitative relationships may eventually become convivial to the point where neither organism exists without the other. Long-term stable symbioses that leads to evolutionary change is called "symbiogenesis." These mergers, long-term biological fusions beginning as symbiosis, are the engine of species evolution.*

One of the forerunners of the view that symbiogenesis could be the engine of speciation and heritable variation is, perhaps paradoxically, Richard Dawkins. In *The Selfish Gene* and *The Extended Phenotype* Dawkins sights many examples of symbiosis, and even claims that the symbiotic origin of cells: "is one of those revolutionary ideas which it takes time to get used to, but it is an idea whose time has come" (Dawkins, 1999a, p.172). Perhaps Dawkins is right – the time for revolution has come; however, this revolution is broader in scope than Dawkins initially realised, for the revolution has developed to the point where it seeks to replace the neo-Darwinian 'modern synthesis'. Dawkins (1999a, pp.181-2) claims that:

Symbiotic relationships of mutual benefit are common among animals and plants. A lichen appears superficially to be an individual plant like any other. But it is really an intimate symbiotic union between a fungus and a green algae. Neither partner could live without the other. If their union had become just a bit more intimate we would no longer have been able to tell that a lichen was a double organism at all. Perhaps then there are other double or multiple organisms which we have not recognized as such. Perhaps even we ourselves?

This is a good statement of how symbiosis can lead to symbiogenesis; the intimate union of different species gives rise to a new species. Symbiogenesis is the 'becoming just a bit more intimate'. We will see that the symbiogenesis theory entails that not just "we ourselves", but that *every* species on the planet is a symbiotic union. Dawkins (1999a, p.182) also claims that: "We are gigantic colonies of symbiotic genes." It is quite strange that Dawkins can claim that genes are symbiotic, cells are symbiotic, that we ourselves are symbiotic organisms, and then not conclude that symbiosis is the dominant force behind variation and speciation. He even realizes that two organisms can fuse into one. In *The Extended Phenotype* Dawkins (1999b, pp.222-3) approvingly cites Smith, who claims that:

> *In the cell habitat, an invading organism can progressively lose pieces of itself, slowly blending into the general background, its former existence betrayed only by some relic. Indeed, one is reminded of Alice in Wonderland's encounter with the Cheshire Cat. As she watched it, "it vanished quite slowly, beginning with the tail, and ending with the grin, which remained sometime after the rest of it had gone."*

Let us now look at the symbiogenesis theory from the perspectives of 'evolution as mechanism' and 'evolution as path'. Earlier in the chapter we noted that Darwin had little to say about the causes of heritable variation, and that according to the 'modern synthesis' the overwhelming majority of this variation comes from random gene mutations. These mutations when they are correlated with geographical separation lead to speciation. In contrast, according to Margulis and Sagan (2002, p.11): "random mutation is wildly overemphasized as a source of hereditary variation". Furthermore, according to Margulis (2001, p.33): "most evolutionary novelty arose and still arises directly from symbiosis." Margulis and Sagan (2002, pp.11-12, p.12) expand these claims in the following passages:

Mutations, genetic changes in live organisms, are induc-
ible; this can be done by X-ray radiation or by addition of
mutagenic chemicals to food. Many ways to induce muta-
tions are known but none lead to new organisms. Mutation
accumulation does not lead to new species or even to new
organs or new tissues. If the egg and a batch of sperm of a
mammal is subjected to mutation, yes, hereditary changes
occur, but...99.9 per cent of the mutations are deleterious.

the major source of inherited variation is not random
mutation. Rather, the important transmitted variation that
leads to evolutionary novelty comes from the acquisition of
genomes. Entire sets of genes, indeed whole organisms each
with its own genome, are acquired and incorporated by
others. The most common route of genome acquisition,
furthermore, is by the process known as symbiogenesis.

According to the symbiogenesis theory it is the acquisition
of genomes, not random mutation, which leads to speciation.
The central mechanism of the 'modern synthesis' is therefore
demoted; Margulis and Sagan (2002, p.29) claim that:
"Although random mutations influenced the course of evolution,
their influence was mainly by loss, alteration, and refinement."
It is useful to go back to the origins of life on Earth in order to
appreciate this fundamental difference between symbiogenesis
and the 'modern synthesis'.

Dawkins (1999a, p.19) postulates that replicators came first
and that the cell originated in their service: "replicators perhaps
discovered how to protect themselves, either chemically, or by
building a physical wall of protein around themselves. This may

have been how the first living cells appeared." According to Dawkins (1999a, p.46) these cells are survival machines which "began as passive receptacles for the genes, providing little more than walls to protect them from the chemical warfare of their rivals and the ravages of accidental molecular bombardment." Clearly, this view of the origins of life – that the cell is protective armour to help win a war – is going to lead to the construction of a 'selfish' story.

In contrast, Margulis (2001, pp.91-2) postulates that the cell membrane originated *first*, and that the replicators came *second*: "cell-like membranous enclosures form as naturally as bubbles when oil is shaken with water...Prelife, with a suitable source of energy inside a greasy membrane, grew chemically complex...After a great deal of metabolic evolution, which I believe occurred inside the self-maintaining greasy membrane, some, those containing phosphate and nucleosides with phosphate attached to them, acquired the ability to replicate more or less accurately." In this account there is no need for protective armour because there is no war; in contrast to the former account it is not assumed that there is a competition-cooperation dichotomy.

It is claimed by Margulis and Sagan that these first bacterial cells marked the origins of life, but that they did not mark the origin of species. For them genetic material is so easily transferred between *all* the bacteria in the world that it is nonsensical to say that there are species of bacteria; speciation is solely a property of nucleated organisms. They are thus able to claim that (2002, pp.55-56): "The creative force of symbiosis produced eukaryotic cells from bacteria. Hence all larger organisms – protoctists, fungi, animals, and plants – originated symbiotically." In other words, as the eukaryotic cell itself is a symbiotic construction (which Dawkins claimed was an idea whose time had come), and as eukaryotic cells are definitive of a species, this means that all species are in origin symbiotic constructions.

This claim by itself is perhaps not that radical – a giraffe is a 'symbiotic' creation because it is composed of symbiotically created cells. However, eukaryotic cell formation is just 'stage one' of the symbiogenesis theory. 'Stage two' concerns all of the evolution of life since this time; Margulis and Sagan (2002, p.56) claim that: "Details abound that support the concept that all visible organisms, plants, animals, and fungi evolved by "body fusion". Fusion at the microscopic level led to genetic integration and formation of ever-more complex individuals."

The focus of the symbiogenesis evolutionary mechanism is the microcosm. Microbes – which have uniquely capable complete genomes – drive evolutionary change forward. According to Margulis and Sagan (2002, p.72) these genomes "come neatly packaged with long histories of heritable virtuos-ities and synthetic tricks. They provide just what is needed for an organism to change drastically and yet remain coherent and viable." Initially there is an association between two genomes, this can lead to a partnership, a symbiosis, and finally to a new species formed by symbiogenesis. While the 'modern synthesis' mutations are *random*, this process is driven by *reasons* (Margulis and Sagan, 2002, p.89):

For two different types of genomes to merge and form a new one, the organisms themselves must have a reason to come together. Reasons vary. Organism A may find B delicious, and try to swallow B. Alternatively, organism A

may require the chemical form of nitrogen excreted in the
waste of B. Organism A may simply bask, at first, in the
shade provided by B – or B may sequester the alkaline
moisture that exudes at dawn from the pores of B. These are
ecological issues with many subtleties, but they underlie the
transfer and eventual merger of microbial genomes to
larger forms of life.

This evolutionary mechanism gives rise to a new and comprehensive definition of a species (Margulis and Sagan, 2002, p.94): "Two live beings belong to the same species when the content and the number of integrated, formerly independent genomes that constitute them are the same." There is a four level hierarchy of 'symbiotic partner integration' which leads to symbiogenetic speciation. The four levels are behavioural integration, metabolic integration, gene-product association, and genetic integration.

The behavioural integration of two symbiotic partners is the most basic and superficial type of symbiotic integration. Ivan Emmanuel Wallin in *Symbionticism and the Origin of Species* named this level of integration "prototaxis" – the innate tendency of one kind of cell or organism to respond in a specific manner to another sort of organism. This means that when two potential partners find themselves in the same place at the same time their behaviour towards each other will be determined by ancestry and contingency. An example is the origin of plants (Margulis and Sagan, 2002, p.99):

No alga or plant ever evolved photosynthesis on its own. All
shared some ancestor – recent or remote – that ate but

failed to digest a green or red or greenish blue bacterial photosynthesizer. Prototaxis, in this case, is the tendency toward hunger on the part of the eater and toward resistance to digestion on the part of the eaten. Starvation in the light and resistance to digestion, in short, have led over and again to permanently pigmented photosynthetic organisms: Algae, lichens, plants, green worms, green hydra, brown corals, and giant clams.

The *metabolic* level is the next most intimate level of symbiotic partner integration. Typically, the metabolic product of one partner is the food of the other; the two-way exchange of these products over time leads to dependency and a loss of individuality. Examples are 'green animals' and lichens. One species of 'green animal' is the flatworm *Convoluta roscoffersis* which is a symbiotic integration of a worm with photosynthetic microorganisms. The photosynthetic products of the microorganisms become food for the worm, and the worm produces nitrogen-rich waste which is required by the microorganisms.

The third level of partner integration is "gene-product association"; proteins or RNA molecules from one partner are required for the functioning of the other. A good example is the symbiotic integration between a pea plant and its nitrogen-fixing bacterial nodules. In the partnership the production of hemoglobin is chimeric – the heme is produced by the bacteria and the globin is produced by the plant. This means that the hemoglobin which is required for the continued survival of both 'partners' is a result of their gene-product integration.

The highest level of symbiotic intimacy is genetic integration. This occurs when a gene of one organism enters and remains with the genes of another; for example, when the gene of a free-living bacterium gets integrated into a plant nucleus.

There will tend to be a movement from the lowest level of integration towards the highest if this is in the interests of both partners; these 'interests' will be determined by natural selection. In some cases behavioural integration will not be transcended by deeper levels of integration, because this level maximises the offspring of the partners. According to Margulis and Sagan (2002, p.102) it follows that: "The complete, irreversible integration of two different beings to form a new one will occur if at all times the physically associated organisms leave more descendents than do the independent unassociated ones." The link between symbiogenesis and natural selection is made clear in the following passage (Margulis and Sagan, 2002, p.102):

> *The hundreds of mitochondria...in each of our cells, never leave these cells. Why? Because the world of animal tissue is full of oxygen and requires a flow of oxygen to the cell at all times. In the oxygen-rich Proterozoic, the cells that retained their mitochondria throughout their lives must have been "naturally selected" over those that occasionally let their mitochondria return to the bacterial world.*

The relationship between symbioses, symbiogenesis, random mutations, and natural selection, should now be clear. Margulis and Sagan (2002, p.157) assert that: "*Protracted symbioses lead to symbiogenesis*: the origin of new organelles, organellar systems, tissues, organs, organisms, and species... *Symbiogenesis*, the inheritance of acquired genomes, mostly those of bacteria and other microbes, is the greatest source of evolutionary innovation...*Random mutations* only refine and alter, but do not produce, species-level change...*Natural selection* directs evolution through propagation and elimination of what it has already."

This view of evolution is backed up by the work of Donald I. Williamson who concludes that there is "a saltational process in animal evolution which operates independently of the accumulation of mutations and then selection" (Margulis and Sagan, 2002, p.165). Williamson's work on invertebrates led him to realise that animal-animal genome transfer is the only thing that can explain the distribution of larval types, as the genome that determines the animal larval form is different from that which determines the adult form. In other words, there have been numerous successful sexual encounters between members of different lineages; Margulis and Sagan (2002, p.166) claim that: "Such successful matings between very distantly related animals occurred infrequently, some thirty to fifty times in 541 million years. This means a fertile, successful outcome happens roughly once in 10 million years."

These encounters occurred by either external or internal fertilization and can, Williamson claims, explain why extremely different adults develop from nearly identical larvae, whereas closely related adults can develop from entirely different larvae. Williamson concludes that (Margulis and Sagan, 2002, p.169): "all species that produce larvae, even caterpillars and other terrestrial animals, acquired foreign genomes at some point in their history." For example, the sergestid shrimp is evolutionarily successful largely because it: "acquired, integrated, and put to work at least four intact genomes" (Margulis and Sagan,

2002, p.169). In contrast, random mutations only have a very minor explanatory role in the success of the shrimp.

The central evolutionary mechanism in the symbiogenesis theory is the acquisition of genomes. When it comes to the higher taxa Margulis and Sagan (2002, p.180) admit that: "The inference of speciation and the evolution of higher taxa by symbiogenesis is often fraught with difficulty." They suggest that mammalian speciation may have occurred through the Todd-Kolwicki Karyotypic Fission Theory. The karyotype details (the total number and the morphology of the entire chromosome set of an organism) that have been collected clearly indicate saltationist speciation. This is because (Margulis and Sagan, 2002, p.192): "If Darwinian gradualism explains the origins of animal and plant species, it follows that closely related species should have similar karyotypes. They don't."

The evidence for the symbiogenetic speciation of mammals is less strong than for the lower taxa. However, it needs to be remembered that the higher taxa only constitute a very small minority of the planet's life-forms. Furthermore, whilst there are no known cases of speciation by random mutation, there are two cases of speciation by symbiogenesis. The first of these is Theodosius Dobzhansky's experiments with *Drosophila*. Dobzhansky bred one group of fruit flies for two years at progressively hotter temperatures, at which point they could not produce fertile offspring with the group that had been breeding at normal temperatures. This was because symbiogenetic speciation had occurred; Margulis (2001, p.9) summarises the findings as follows: "the hot-breeding flies lacked an intracellular symbiotic bacteria found in the cold breeders."

The second case of symbiotic speciation is Kwang Jeon's experiments with amoebae. Following an accident where some new amoebae infected his amoebae collection with a rod-shaped bacterium, most of his collection died. However, a few survived with the bacteria inside them. According to Margulis and Sagan (1997, pp. 121-2) these amoebas: "were easily killed by antibiotics, which, while deadly to bacteria, did not harm his normal "nonbacterized" amoebae. A change was occurring. The two types of organisms, bacteria and amoebae, were becoming one." By switching nuclei between normal and infected cells Jeon was able to show that the infected nuclei needed a bacteria infected cell in order to survive.

It is time to turn to 'evolution as path'; it is clear that symbiogenesis involves a very different view to the 'modern synthesis'. The two main areas of difference are the shape of the phylogenetic evolutionary tree and the presence of 'saltationism' in the evolutionary tree. Recall that the 'modern synthesis' entails that the phylogenetic tree forever branches from a common ancestor. Contrarily, for the symbiogenesist (Margulis and Sagan, 2002, p.15): "the acquisition of heritable genomes can be depicted as an anastomosis, a fusing of branches." Margulis and Sagan (2002, p.172) describe this process of anastomosis quite graphically:

Animal evolution resembles the evolution of machines, where the typewriters and televisionlike screens integrate to form laptops, and internal combustion engines and carriages merge to form automobiles. The principle stays the same: Well-honed parts integrate into startling new wholes.

The second area of difference is 'saltationism' because symbiogenesis is clearly a saltationist theory of evolution. Recall that Dawkins argues against saltationism on the basis of the overwhelmingly likely deleterious effects of a *macromutation*. However, these arguments clearly do not apply to saltationism initiated by *genome acquisition*. Not surprisingly, Margulis and Sagan (2002, p.96) appeal to the jerky nature of the fossil record as support for symbiogenetic saltationism: "Punctuated equilibrium is there for all who take the time to see it. The discontinuous record of past life shows clearly that the transition from one species to another occurs in discrete jumps. In trilobites, snails, seed ferns, horses, lungfish, sharks, and clams, evidence abounds for punctuated change."

We saw that Dawkins claims that punctuated equilibrium is just about *speeds* of evolution, not about saltationism. He also argues that the fossil record only reveals 'migrational events' and not 'evolutionary events'. In contrast, Margulis and Sagan (2002, p.83) cite Niles Eldredge's work with the well preserved Cambrian trilobites:

> *one species would continue with minor random variations for 800,000 years. Another would abruptly begin and overlie the first for 1.3 million years. The search for intermediate forms and gradual evolutionary change between the two species was always futile. The sedimentary rocks in which the glorious fossil record is embedded do not lie. They do not deceive. The record was punctuated, and the differences between species of extinct animals trapped in it were clean and distinct.*

There are three interpretations that can be made of the fossil record. Firstly, one can share the Darwinian belief that the intermediates are missing. Secondly, one can agree with Dawkins that the fossil record only records a 'migrational event'. Thirdly, one can believe that there were no intermediates and that the fossil record reveals a symbiogenetic speciation. For now, we can at least conclude that the fossil record is concordant with the symbiogenesis theory, and its claim that evolution is punctuated, saltational, and driven by microbial mergers. Margulis and Sagan (2002, p.201) claim that: "The reliance on accumulation of random mutations in DNA is not so much "wrong" as oversimplified and incomplete: It misses the symbiotic forest for the genetic trees." Similar sentiments on the importance of random mutations in evolution are expressed by the developmental systems theorists.

2.3 Developmental systems theory

In the 'modern synthesis' the mechanism of evolution is natural selection acting on genes; genes are singled out because they are believed to be the only long-term persisting elements of organisms, and because they are believed to contain 'information' which leads to the development of favourable phenotypic traits. The symbiogenetic theory rejects this view of evolution because it denies that natural selection is the dominant mechanism of evolution; indeed, according to this view natural selection does not give rise to speciation. In contrast, developmental systems theorists reject the 'modern synthesis' because they believe that natural selection acts on the whole life-cycle of an organism; according to this view it is wrong to single out genes as having more causal influence than any other factor that affects an organism.

By not giving priority to any factor in the developmental and evolutionary process developmental systems theorists claim to have dissolved several long-standing dichotomies: the innate/acquired distinction, internal versus external causation, and the opposition between 'nature' and 'nurture'. They claim that the 'modern synthesis' is preformationist, as is any paradigm that accepts such dichotomies. Developmental systems theorists Susan Oyama, Paul E. Griffiths and Russell D. Gray (2001, p.1) pose the question: "Can we shape our destiny, or are we robots programmed by our selfish genes?" This question mistakenly sets up a false dichotomy; for, we have already seen that 'selfish' genes do not imply that organisms are programmed robots devoid of free will. The real point they are trying to make is better expressed in the following passage (2001, p.1):

> *The standard response to nature/nurture oppositions is the homily that nowadays everyone is an interactionist: All phenotypes are the joint product of genes and environment. According to one version of this conventional "interactionist" position, the real debate should not be about whether a particular trait is due to nature or nurture, but rather how much each "influences" the trait.*

So, their objection to the selfish gene theory is not that this view entails genetic determinism. The real objection is that the 'modern synthesis', whilst accepting that environmental factors and genetic factors interact to produce evolutionary outcomes, still sees the genes as containing information *that will lead to a given outcome* if the background environmental resources are available. This is even clearer in the following passage from Oyama (2001, p.179): "Some might say, for instance, that the nature/nurture opposition is nonsensical because some things don't just mature, but require interaction. Here it is evident that *interaction*, far from challenging the concept of internally driven maturation, assumes it."

So, Oyama claims that it is not appropriate to describe genes as "containing information" – information is the result of ontogeny rather than its cause. According to this view the whole developmental process is a life cycle that contains a diverse range of both heritable and non-inheritable resources, and these resources are reconstructed in every cycle through "self-organization". It follows from this that the 'modern synthesis' distinction between replicators and vehicles is inappropriate. This is explained by Bruce H. Weber and David J. Depew (2001, p.240):

> *DST champions infer from the presumptive causal parity of all developmental resources that the replicator/interactor distinction, on which the units of selection debate has been predicated, is ill conceived. When phenotypic traits are construed as developmental resources it can be seen that they are as much replicators as interactors.*

A central reason for adopting developmental systems theory is the belief that a whole range of resources are heritable; the 'modern synthesis' with its sole focus on genes ignores these other resources and holds that they are not relevant to evolution. I will now outline this expanded range of heritable resources and explain why developmental systems theorists believe that they play an important role in evolution. Eva Jablonka (2001, p.99) claims that:

According to genic neo-Darwinism, nucleic acids are the sole units of heritable variation, the transmission of these units is independent of their expression, and the generation of genetic variations is not adaptively guided by the selective environment or the developmental history of the organism. This replicator-centered, gene-derived view of heredity is, however, not only severely limited but also severely misleading. There are multiple inheritance systems, with several modes of transmission for each system, that have different properties and that interact with each other.

Jablonka separates these inheritance systems into four groups: genetic, epigenetic, behavioural, and symbolic. Because these systems are continuously interacting with each other this means that the unit of evolutionary selection has to be the organism as a whole.[5] I will now outline Jablonka's four groups of inheritance systems.

The genetic inheritance system simply refers to genes; Jablonka (2001, p.100) states that: "The gene is a template made up of nucleotides whose sequential organization can be transformed through a complex process of decoding into functional RNA and proteins." The behavioural inheritance system includes inducing-substance transfer; Jablonka (2001, p.110) claims that: "mammal fetuses are able to smell semivolatile liquids transferred to them across the mother's placenta,

---

[5] Jablonka's views are very similar to developmental systems theory; the difference arises because Jablonka does not claim that *all* of the inheritance systems are wholly devoid of information.

and later show preference or aversion for food items carrying these smells." Another element of the behavioural inheritance system is imitation and non-imitative social learning. The symbolic inheritance system also entails social learning and imitation, the difference is that the information here is encoded; according to Jablonka (2001, p.112) this gives symbolic inheritance: "unlimited heredity and huge evolutionary potential."

The fourth and final system is the epigenetic inheritance system; this system relates to cellular inheritance between generations, and includes the *steady-state system, structural inheritance* and *chromatin-marking*. In contrast to genetic inheritance, epigenetic inheritance is generally patterned and holistic. The *steady-state system* of inheritance is a self-perpetuating system of gene product activity within a cell which is initiated by environmental stimuli. Its presence means that genetically identical cells will be heritably different if their ancestor cells have a different developmental history. Jablonka (2001, p.105) claims that these different cellular states: "may be practically unlimited, and cumulative evolutionary change may occur." *Structural inheritance* is simply the templating of existing cellular structures to form new structures. Finally, *chromatin-marks* affect gene expression; this means that genetically identical cells can have heritable differences that can be environmentally induced. Jablonka (2001, p. 106) states that: "The type, the density, and the pattern of marks on a chromosome region affect its potential transcriptional state, and changes in marks can be induced by the changes in the environment."

Griffiths and Gray give an example of how these non-genetic inheritance systems can give rise to fitness differences and adaptation though natural selection. This example involves various colonies of the North American fire ant *Solenopsis invicta*. There are no significant genetic differences between these colonies, but some have large monogynous queens, and others have small polygynous queens. The way that a particular queen develops depends entirely on the colony in which it is raised – exposure of eggs from a monogynous colony to the pheremonal "culture" of a polygynous colony produces small queens who found further polygynous colonies. According to Griffiths and Gray (2001, p.199) this means that: "a mutation in a nongenetic element of the developmental matrix can induce a new self-replicating variant of the system which may differ in fitness from the original."

Utilizing these diverse inheritance systems Griffiths and Gray argue for a developmental systems framework of natural selection that can replace the gene-centered version. In their revised terms (2001, p.214): "Developmental system – the interactions and processes that produce a life cycle...Natural selection – the differential reproduction of heritable variants of developmental systems due to relative improvements in their functioning... Adaptation – the product of natural selection...Evolution – change over time in the composition of populations of developmental systems." Griffiths and Gray (2001, p.200) argue that this framework widens the scope of Darwinian natural selection to more closely reflect reality:

Not only might expanded forms of inheritance play an important role in the generation of evolutionary novelty, they could also significantly alter the dynamics of evolutionary change...expanded inheritance can facilitate transitions from suboptimal to higher peaks, thus creating more effective evolutionary dynamics than would be possible under strict genes-only conceptions. Expanded forms of inheritance may also be the cause of reproductive isolation and hence of speciation.

Let us recap. According to developmental systems theory the life-cycle of an organism is constructed through joint-determination by multiple causes. Some of these causes are non-genetic but internal to an organism; these causes are disregarded by the 'modern synthesis' as unimportant. Other causes result from the interaction between an organism and its surroundings; these causes the 'modern synthesis' considers to be a separate and isolated set of processes to those that go on inside the organism. From the holistic perspective of developmental systems theory this separation is an unwarranted and an unhelpful one. According to this perspective no causal factor is in control of the developmental process, and this is because every factor has its effects only in conjunction with feedback effects from the output of other factors. As Oyama (2001, p.182) puts it:

the claim is hardly that genetic effects on organisms cannot be identified, but that the genes have their effects by being affected by other factors – by their cellular environments, if you will – and these often include the very processes they influence. The impact of gene products, furthermore, tends to vary with other conditions. Starting an account with genetic transcription, and treating the DNA as an "independent variable" that "initiates" an interesting cascade of events, leads only too easily to obliterating from the causal landscape the events and conditions that brought that tran-scription about.

When one appreciates this joint reciprocal causation between factors then one can also appreciate another aspect of developmental systems theory. The 'modern synthesis' presumes that environmental niches exist independently of organisms, and that it is random mutations in organisms which lead to organisms 'fitting' these niches; contrarily, according to developmental systems theory organisms create their own niches. Niches are created both by the organism itself and by its ancestors. Oyama, Griffiths and Gray (2001, p.6) claim that: "there are no preexisting niches or environmental problems that shape populations from without." They explain this more fully in the following passage (2001, p.6):

If evolution is change in developmental systems...it is no longer possible to think of evolution as the shaping of the organism to fit an environmental niche. Rather, the various elements of the developmental systems coevolve. Organisms construct their niches both straightforwardly by physically transforming their surroundings and, equally importantly, by changing which elements of the external environment are part of the developmental system and thus able to influence the evolutionary process in that lineage.

Another advocate of organism-created niches is Richard Lewontin. Lewontin (2000, p.49) claims that: "The concept of an empty ecological niche cannot be made concrete. There is a non-countable infinity of ways in which the physical world can be put together to describe an ecological niche, nearly all of which would seem absurd or arbitrary because we have never seen an organism occupying such a niche." Furthermore, Lewontin (2000, p.66) asks: "Are there any circumstances in which it can be said that organisms "adapt" to an externally imposed environment rather than "constructing" it by their life activities?" Lewontin (2000, p.67) concludes that there aren't because "there are no environments without organisms."

How does an organism construct its environment? Lewontin claims that organisms determine which aspects of their surroundings are relevant to them, they actively construct a world around themselves through creating a surrounding atmosphere, and they also continuously alter their surroundings. Furthermore, organisms can modulate the statistical properties of external conditions, and their biology determines the actual physical nature of the signals that arrive from their surroundings (Lewontin, 2000, pp.51-63).

To recapitulate, the central claim of developmental systems theory is that the elevation of genes to a prominent role in development is unwarranted. The elevation of genes in the 'modern synthesis' is justified by reference to the Central Dogma of molecular biology which claims that information in DNA and RNA codes for proteins, but that information can never run directly from proteins back to DNA and RNA. The developmental systems theorists reject this view because they believe that information is not something that pre-exists the biological interactions with which it is associated. For example, the zygote contains DNA sequences but it is not determined in advance precisely which proteins these sequences will code for. As Peter Godfrey-Smith (2001, p.293) explains: "The DNA exists in the zygote, but the eventual protein products of that DNA depend on a great array of other components, many of which will only be built up in the course of development."

Let us now consider 'evolution as path'. What we would expect to find in the fossil record if the developmental systems theorists are correct? The environment plays such a big role in the construction of the life-cycle of an organism that, *ceteris paribus*, one would expect that an unchanging environment would not generate much variation and speciation. Contrarily, a moderately changing environment would be expected to lead to a higher level of variation and speciation, and a massive environmental perturbation would be expected to lead to a vast amount of variation and relatively rapid speciation.

These expectations are clearly concordant with the theory of punctuated equilibrium; we certainly wouldn't expect to see an equal speed of transition between species if developmental systems theory is correct. What we would expect is a period of 'equilibrium' wherein organisms are incorporating minor environmental adjustments into their life-cycles, followed by a period of 'punctuation' when big environmental changes occur. We have seen that the symbiogenesis advocates give a seemingly different explanation – they claim that the punctuated periods are due to symbiogenesis. These two explanations of punctuation may actually be concordant, for the symbioses that lead to symbiogenesis are a product of environment-induced 'prototaxis'; and there will clearly be far more 'prototaxis' events in a period of big environmental change. Direct environment-induced organismic change, which is heritable, is a feature of both of these paradigms, and it is a feature which is missing from the 'modern synthesis'.

2.4 The state of the modern synthesis

The main weakness of modern evolutionary theory is its lack of a fully worked out theory of variation, that is, of candidature for evolution, of the form in which genetic variants are proffered for selection. We have therefore no convincing account of evolutionary progress – of the otherwise inexplicable tendency of organisms to adopt ever more complicated solutions of the problems of remaining alive.

Sir Peter Medawar
(1967 cited in Dawkins, 1999b, p.165)

We have seen that there are many reasons to doubt that we understand how the evolutionary process works. We first saw that this was so at the very start of this chapter where we encountered Simon Conway Morris's summary of the current situation. This is not a new realization as is clear from Medawar's claim, in 1967, that we have "no convincing account of evolutionary progress". A key weakness in the neo-Darwinian 'modern synthesis' is its failure to give a convincing account of Darwin's missing variation.

I have outlined two contemporary theories which seek to explain the sources of heritable variation without sole reliance on random mutations. According to developmental systems theory there are a wide range of non-genetic heritable factors that create evolutionary variation, and it is the whole life cycle that replicates; as natural selection operates on outcomes, the replicator-vehicle dichotomy becomes meaningless. In contrast, according to symbiogenesis theory the overwhelming majority of variation comes from genome acquisition via symbiogenetic fusions; natural selection works on both the species formed by symbiogenesis and on the trivial variation caused by random mutations, and thereby provides *fine-grained* adaptation.

Do these two theories pose a serious challenge to the 'modern synthesis'? When it comes to the *path of evolution* it is seemingly very hard to pose a serious challenge to the 'modern synthesis' view. Dawkins argues that the fossil record in one place only reveals a 'migration event' rather than an 'evolution event'; this means that we would not expect to find any intermediates. This implies either that the intermediates could be found elsewhere, or that we fall back onto the traditional Darwinian claim that the intermediates are missing. So the intermediates are either undiscovered or missing. This is hardly convincing stuff – but it is also irrefutable; *whatever* the fossil record is like an advocate of the 'modern synthesis' can claim that it supports their theory.

The greater responsiveness of organisms to environmental influences which is hypothesized in developmental systems theory can, perhaps, more adequately explain punctuated equilibrium; this is due to the different cycles of environmental change that occur over geological timescales. Symbiogenesists claim that the punctuations reflect true saltationism – in the form of the merging of formerly independent organisms to create a new one. As symbioses originate from behavioural "prototaxis" they would escalate in times of environmental upheaval, and so could also explain the differential speeds inherent to punctuated equilibrium. If evidence continues to accumulate which supports both the symbiotic origins of the eukaryotic cell and symbiogenetic speciation (as we saw with *Drosophila* and amoebae), then the case for a saltational evolutionary path which encapsulates both branching and fusing phylogenies will become stronger.

The debate over 'evolution as mechanism' whilst appearing to involve three irreconcilable positions, actually appears to be mainly concerned with relative weightings. All three positions accept that random mutations and natural selection play a role in evolution. Furthermore, all three positions accept that mechanisms other than natural selection and random mutations play a role in evolution. Dawkins (2006, p.318) admits that:

> *If there are views of the evolution theory that deny slow gradualism, and deny the central role of natural selection, they may be true in particular cases. But they cannot be the whole truth, for they deny the very heart of the evolution theory, which gives it the power to dissolve astronomical improbabilities and explain prodigies of apparent miracle.*

It seems that none of the three paradigms in their extreme form is the 'whole truth'. The question is what the relative weighting should be between the following: symbiotic saltationism, expanded inheritance, environmental determination of genetic information, natural selection, and random mutations. Dawkins argues that the overwhelmingly dominant force in the evolutionary mechanism is the natural selection of random genetic mutations, other factors are of minor significance; he accepts that the environment is important in development, but believes that genes contain information that has a statistical probability of leading to a given trait. In contrast, developmental systems theorists argue that there is no genetic information without an environment; this means that, in effect, it is the multitude of interacting and heritable environmental factors – which play a large part in forging the life-cycles which get naturally selected – that need to be given the dominant role in the evolutionary mechanism. Finally, the symbiogenesists reject the idea that the dominant force is natural selection, of either genes, or life-cycles, and claim that the dominant evolutionary mechanism is genome acquisition.

So, do the two challenging theories make a compelling case for taking the dominant role? Margulis and Sagan make a strong case for increasing the weighting of symbiogenesis in evolution. However, it is possible that the claims that they make are slightly too strong. They take a hard line on mutations by claiming that they never lead to speciation – they just alter and refine. Margulis and Sagan (2002, p.72) claim that: "mutations do not create new species." This claim rests entirely on their definition of a species. David L. Hull (1998, p.297) claims that: "Species are anything that anyone chooses to make them." This means that an alternative definition of a species – such as 'recognition, mating, and production of fertile offspring' – could lead to the conclusion that an 'alteration' of one species through accumulated mutations actually leads to the creation of a new species. This would be particularly likely in a small isolated island population which would have a high level of inbreeding, and which would be isolated from the gene flow of the rest of the species. So, whilst symbiogenesis could be the dominant form of speciation, it is seemingly possible that there could also be speciation without genome acquisition.

The developmental systems theorists also make a strong case for taking the unit of selection to be the organism as a whole, and for abandoning the replicator-vehicle dichotomy. There is surely a large role in generating variation and adaptation played by non-genetic heritable factors. Furthermore, the claim that there is no protein-coding information in DNA that is environment-independent could well turn out to be true. The idea that organisms are jointly determined by multiple causes in a process of self-organization looks set to become increasingly dominant.

I would suggest that a strong case can be made that the dominant mechanism in evolution is actually a synthesis of symbiogenesis and developmental systems theory. In the symbiogenesis theory natural selection is utilized in a minor capacity to explain the fine-grained adaptiveness of organisms to their environmental niches; the overwhelming majority of variation, and all speciation, comes from symbiogenesis. In developmental systems theory there are no environmental niches that pre-date organisms, as organisms create their own niches.

By combining these two ideas we can postulate an adequate explanation for adaptiveness with no role for positive selection. It is symbiogenesis that generates the overwhelming majority of variation and speciation, and the resulting organisms then create their own niches. In this way we can explain speciation and adaptiveness without positive natural selection. This view is the actual fulfilment of the wishes of H. J. Muller (1940 cited in Ruse, 1989, p.90) who claimed that: "if selection could be somehow dispensed with, so that all variants survived and multiplied, the higher forms would nevertheless have arisen."

It can be concluded that the 'modern synthesis' explanations for both the path and the mechanism of evolution are inadequate. Nevertheless, it could be argued that all three of the theories that I have outlined can be synthesized into one overarching theory. Heritable variation would be generated by symbiogenesis, expanded inheritance and random mutations; whilst, speciation would also occur through symbiogenesis, random mutation accumulation and expanded inheritance.

In this overarching theory there would be no organisms, or information, without an environment; but given that there is a species-created environment, there is a sense in which the genes that relate to that environment could be described as 'selfish'. This is because a species-created niche creates a high statistical probability that a given gene will be both created and lead to a given trait – this is a 'selfish' attempt to preserve the niche. In concordance with this, we concluded that Dawkins has to fall back onto an equivalence of genotypic with phenotypic traits which is highly contingent on the environmental history of a particular organism – a history which, in this overarching theory, is created by the organism itself. Furthermore, extended phenotypic effects can be seen as simply a small part of the plethora of organism-environment interactions that constitute the replication of an organism's life-cycle. So, rather than any one of the three theories being dominant, it seems that each can give an important perspective on the larger evolutionary picture.

2.5 Non-biological evolution

In the previous sections we have focused solely on biological evolution – the mechanisms of biological evolution, and the paths that biological evolution has taken. In other words, we have focused on life-forms, and how these life-forms have brought forth other life-forms. If one fully embraces the evolutionary paradigm then one will trace the evolutionary process back to the fireball that was the Big Bang. One will talk of the evolution of minerals, the evolution of planets and the evolution of solar systems; we can reasonably assume that evolution was occurring a very long time before the evolution of biological evolution.[6]

Compared to life-forms, planets and solar systems are vastly larger in size. However, the mechanisms which underpin the formation of planets and solar systems are still at work in biological evolution. It is interactions between the *smallest* constituents of the universe which gives the impetus to the evolution of the universe. The smallest constituents – which on one popular contemporary theory are tiny vibrating 'strings' – give rise to atoms, which, in turn, interact in such a way as to form molecules. As these interactions 'scale up' we end up with biological life-forms and planets. It is the interactions at the lowest level, and the mechanisms of evolution which are operating at this level, which brought forth and underpin

---

[6] It is an interesting question whether there is distinction in the universe itself between the 'biological' and the 'non-biological'; it is possible that this distinction solely exists in the human conceptual realm. It seems obvious to me that there is a distinction in the universe itself between the 'biological' and the 'non-biological'; however, it also seems obvious that this distinction is a subtle one rather than a chasm. One could think of the 'biological' part of the universe as a state of 'excitation' or 'exhilaration'. There is clearly a distinction between the states of the universe which exist when one is 'exhilarated' compared to the states of the universe which exist when one is 'depressed'; furthermore, a state of 'exhilaration' is a better state of the universe than a state of 'depression'. Analogously, a state of 'biological' is a better state of the universe than a state of 'non-biological'. Whilst there is a clear distinction between 'exhilaration' and 'depression', and between 'biological' and 'non-biological', these distinctions do not entail the existence of chasms between the parts of the universe that instantiate them. Furthermore, given the subtleness of the distinction there are good grounds for believing that 'biological evolution' is fundamentally similar to 'non-biological evolution'.

evolution at the biological level. In short, the mechanisms of evolution which resulted in the evolution of minerals, planets, solar systems and biological life-forms, are still in existence today; furthermore, these mechanisms exist today *within* biological life-forms.

Can we fully understand the nature of these non-biological evolutionary mechanisms? This question is equivalent to this question: Can we fully understand the nature of our surroundings? It seems obvious that we cannot fully understand the nature of these mechanisms; this is because we gain access to our surroundings via our senses. And our senses only give us access to the exterior of things – not their interior. If we could see the exterior of the smallest constituents of our surroundings (strings?) we would still not know their interior – we would not know if there is anything it is like to be a string; we also would not know why a string forms the atom that it does, or why it moves the way that it does. All we can do is observe from without and create terms such as 'gravity', 'EM force', and 'strong force', in an attempt to understand what is going on.

If we don't fully understand the mechanisms of evolution at this fundamental level, then we cannot fully grasp the forces which underpin biological evolution. And, even if we were to focus solely on the 'upper' biological level (one can think of life-forms as containing two interlinked evolutionary mechanisms – the upper biological level and the lower more fundamental non-biological level), then we would still be barred from the interior side of the processes which are occurring. Many philosophers have realised that we cannot fully understand the nature of our surroundings. For example, here are some assertions by Immanuel Kant, Peter Unger, Bertrand Russell, and Galen Strawson:

For every substance, including even a simple element of matter, must after all have some kind of inner activity as the ground of its producing an external effect, and that in spite of the fact that I cannot specify in what that inner activity consists...Leibniz said that this inner ground of all its external relations and their changes was a power of representation. This thought, which was not developed by Leibniz, was greeted with laughter by later philosophers. They would, however, have been better advised to have first considered the question whether a substance, such as a simple part of matter, would be possible in the complete absence of any inner state.

(Kant, 1992, p. 315)

Except for what little of the physical world we might apprehend in conscious experience, which is available if Materialism should be true, *the physical is mysterious to us*.

(Unger, 2006, p.5)

we know nothing about the intrinsic quality of physical events except when these are mental events that we directly experience.

(Russell, 1956, p.153)

I take physicalism to be the view that every real, concrete phenomenon in the universe is ... physical...I will equate 'concrete' with 'spatio-temporally (or at least temporally) located', and I will use 'phenomenon' as a completely general word for any sort of existent. Plainly all mental goings on are concrete phenomena... But how can experiential phenomena be physical phenomena? Many take this claim to be profoundly problematic (this is the 'mind-body problem'). This is usually because they think they know a lot about the nature of the physical. They take the idea that the experiential is physical to be profoundly problematic *given what we know about the nature of the physical*. But they have already made a large and fatal mistake. This is because we have no good reason to think that we know anything about the physical that gives us any reason to find any problem in the idea that experiential phenomena are physical phenomena.

(Strawson, 2006, pp.3-4)

In short, a simple part of matter has an inner state (Kant), the physical is mysterious (Unger), we know nothing about the intrinsic quality of the overwhelming majority of physical events (Russell), and these events could be experiential events (Strawson). According to this view, which I take to be obviously correct, humans are incapable of knowing the fundamental nature of the universe. If this is right, then humans won't be able to fully understand the evolutionary process. Humans would be unable to fully comprehend the nature of their surroundings and why they evolve the way that they do; this applies to both 'biological evolution' and 'non-biological evolution'. In the next section we will consider the opposing view – the view that humans are able to access the interior states of their surroundings.

2.6 The view that humans can access the interior states of their surroundings

In opposition to the view which I outlined in the previous section there are those who believe that humans are able to access the interior states of their surroundings. In this section I will explore and reject this view. The favourite example of the supporter of this view is that a human can access the inner state of other humans that is the feeling of pain, so let us consider this example. The two opposing views are clear in the following quotes:

> when one speaks about "one's own physical pain" and about "another person's physical pain," one might also appear to be speaking about two wholly distinct orders of events...So, for the person in pain, so incontestably and unnegotiably present is it that "having pain" may come to be thought of as the most vibrant example of what it is to "have certainty," while for the other person it is so elusive that "hearing about pain" may exist as the primary model of what it is "to have doubt." Thus pain comes unsharably into our midst as at once that which cannot be denied and that which cannot be confirmed.

> (Scarry, 1985, p.4)

> In what sense are my sensations *private*? – Well, only I can know whether I am really in pain; another person can only surmise it. – In one way this is wrong, and in another non-sense. If we are using the word "to know" as it is normally

used (and how else are we to use it?), then other people very often know when I am in pain.

(Wittgenstein, 1988, p.89)

But do we always distinguish between 'mere behavior' and experience + behavior'? If we see someone falling into flames and crying out, do we say to ourselves: "there are of course two cases: . . ."?

(Wittgenstein, 1968)

If I see someone writhing in pain with evident cause I do not think: all the same, his feelings are hidden from me.

(Wittgenstein, 2006, p.204)

Elaine Scarry makes the case that a human is unable to access the feeling states of pain that are interior to other humans, and Ludwig Wittgenstein makes the case that such access is possible. The first thing to note is how weak Wittgenstein's claims are. He claims that other people "very often know" when he contains feeling states of pain; this means that people *very often do not know* when he contains feeling states of pain. That this is so is clear from his claim that he only thinks that he can access the pain of other people when there is "evident cause"; when there is no such cause he has no such access. This is also revealed in the second Wittgenstein quote above which starts: "But do we always distinguish"; clearly *sometimes we do* distinguish the existence of feeling states which exist within other humans and their behaviour. Furthermore, Wittgenstein is clearly not claiming that a human can *fully* access the interior states of other humans; he is simply postulating that a human can access *some of* the interior states of other humans in some circumstances.

On Wittgenstein's' view it is the observation of another human's behaviour that occasionally leads one to the belief that one is observing their interior feeling states. If one observes Wittgenstein jumping into a blazing fire and hears him crying out as he lands in the flames, then one *knows* that Wittgenstein contains feeling states of pain. What are we to make of such a claim? In such a scenario it seems natural to believe that Wittgenstein contains feelings states of pain; one might even assert that one can see that Wittgenstein is in pain, that one knows that he is in pain. However, accessing the feeling states of pain which are interior to Wittgenstein requires much more than this. Such a feat of inner access requires one being able to access the phenomenal character of Wittgenstein's pain, its qualitative nature, the way that it feels. One clearly cannot achieve such a feat with one's eyes and one's ears, and there are seemingly no other perceptual connections that exist which would enable one to access the inner feeling states of Wittgenstein when he lands in the fire.

If there is no perceptual connection which would enable a human to access the phenomenal character of the states which exist within other humans, then it seems that we should reject Wittgenstein's view. Instead we should agree with William James (1912, p.77):

I see your body acting in a certain way. Its gestures, facial movements, words and conduct generally, are 'expressive,' so I deem it actuated as my own is, by an inner life like mine.

Let us conclude that a human has no perceptual access to the inner states of another human; one has no access to the phenomenal character of the states which exist within them. And, even if we accepted that such a feat of inner access was possible, we have seen that Wittgenstein's claims are so weak that it would still be the case that humans are unable to access the vast majority of states which are interior to other humans. This conclusion has wider implications – for, if one is unable to access all, or the vast majority of, the states which are interior to other humans, then one is also unable to perform such a feat of inner access when it comes to all of one's non-human surroundings.

2.7 Conclusion

Humans do understand some things about evolution. Our senses give us access to certain parts of the universe and this has enabled us to gain knowledge of some of the things which are going on within life-forms; so, we can be fairly confident that natural selection, developmental systems theory and symbiogenesis all have a role to play in biological evolution. We have also gained knowledge of the non-biological evolutionary processes which preceded the evolution of biological life-forms and which still continue today.

However, assuming that we cannot fully understand the nature of our surroundings, because we are barred from accessing the interior states of those surroundings, we will forever lack the knowledge that would enable us to fully understand the nature of both biological evolution and non-biological evolution. Of course, we can often make accurate predictions (scientific or otherwise) but this doesn't mean that we fully understand the nature of the phenomena that we are successfully predicting.

The universe, non-biological evolution and biological evolution are all fundamentally mysterious to us, and will remain so in the future. Our understanding of the mechanisms and paths of evolution seems certain to increase; however, our knowledge will always be limited.

Chapter 3

Conceptions of one's surroundings

In *Sections 2.5 – 2.6* I explored and defended the idea that a human is unable to access the interior states of their surroundings. This inability means that humans cannot fully understand the nature of their surroundings. Nevertheless, humans have a particular conception of their surroundings. This conception has two aspects: firstly, the belief that one's surroundings contain *objects* such as trees, frogs, planets, tables and other humans; secondly, the belief that one's surroundings *in general* are a particular way (that they are free *or* determined, minded *or* matter, similar to one *or* radically different to one, etc.). The *general* conception that one has of one's surroundings is clearly parasitic on the belief that one has concerning the particular objects that exist in one's surroundings. The interactions that one has with the particular objects that one believes to exist in one's surroundings leads to the formation of either an implicitly assumed, or a rationally-derived, general conception of those surroundings. My aim in this chapter is to explore how the particular conception of their surroundings that a particular human has arises from the relationship that exists between that human and their surroundings.

Human surroundings clearly have a particular degree of similarity to humans (at the extremes they are *either* very similar *or* very dissimilar). That is to say, humans have a particular range of attributes *and* human surroundings have a particular range of attributes. However, there are a multitude of ways in which humans can, in general, conceive of their surroundings in relation to themselves. That is to say, human surroundings have a particular bundle of attributes – Bundle S* – but there are a plethora of different beliefs which humans can possibly have concerning which bundle of attributes their surroundings have – Bundle A*, Bundle B*, Bundle C*, Bundle Z*, etc. At the extremes a human could conceive of their surroundings as exceptionally similar to themselves or as exceptionally dissimilar to themselves. In other words, a human could believe that their surroundings have the vast majority (or all) of the attributes that they have *or* they could believe that the vast majority of their surroundings have barely any of the attributes that they have. Why is this important? It is important because it means that there will be a plethora of different views concerning the nature of human surroundings.

In *Section 3.1* I consider whether the way in which a human gains information about their surroundings is inevitably constrained by their perceptual apparatus. In *Section 3.2* I defend the view that human perceptual connections connect to the 'surroundings in-themselves'; I also claim that the 'surroundings in-themselves' are a 'blobject', that perception is non-conceptual, and that objects do not exist in the absence of a perceiver. Then, in *Section 3.3,* I consider the *general* conception that a human has of their surroundings. In *Section 3.4* I explore the relationship between a human's perception of their surroundings and their conception of those surroundings. These sections set the scene for *Section 3.5* in which I consider various views concerning how similar humans are to their surroundings. Finally, in *Section 3.6,* I draw some conclusions.

3.1 Is a human's perceptual apparatus inevitably constrained?

One of the most fundamental aspects of the relationship between humans and their surroundings is the process by which humans perceive their surroundings. The processes of visual perception, auditory perception, olfactory perception, gustatory perception and tactile perception are processes which provide a connection between humans and their surroundings.

The purpose of this section is to consider the nature of these perceptual connections. Do these connections provide access to objects which exist in the absence of the connections? Or, do the connections themselves bring the objects into existence? Alternatively, are both of these possibilities wrong? If the connections bring the objects into existence this implies that a human's perceptual apparatus is inevitably constrained – a human would have no perceptual access to the 'surroundings in-themselves', they would only have access to the objects which are created by their perceptual apparatus. Furthermore, even if these connections provide access to objects which exist in the 'surroundings in-themselves', they might only connect with a very small segment of these surroundings. If this is so, then a human's perceptual apparatus would still be inevitably constrained. This is important because if an inevitable perceptual constraint exists this means that humans are likely to have an inaccurate *general conception* of the universe; for their typical thoughts will either not be about the 'surroundings in-themselves', or they will be about a very small segment of the 'surroundings in-themselves'.[7]

---

[7] By 'typical thoughts' I mean the thoughts that most humans have most of the time concerning their surroundings. If one gets really philosophical and starts

In *Sections 3.1.1 – 3.1.4* I will assume that the perceptual connections of a human connect to the 'surroundings in-themselves'. The conclusion of these sections is that even if such a connection exists the human perceptual apparatus is inevitably constrained in several ways. Then, in *Section 3.2* I will explore and defend the idea that the perceptual connections of a human connect to the 'surroundings in-themselves'; I will also defend the idea that there are no objects in the 'surroundings in-themselves'.

3.1.1 *Constraint versus Inevitable Constraint*

It is useful to distinguish a 'constraint' from an 'inevitable constraint'. It is seemingly exceedingly obvious that human perceptual connections are constrained. After all, it is well known that other animals, such as dogs, can perceive sounds that the human perceptual apparatus is unable to perceive. This means that there are parts of human surroundings which humans are unable to perceive because they don't have the required perceptual connections. Is this constraint an inevitable constraint? One could reasonably contend that this type of constraint is not an inevitable one because it is theoretically possible that technological implants could open up new connections; if such implants were put into a human ear they might be able to perceive exactly the same range of sounds as a dog. Let us grant that this is possible; nevertheless, this alerts us to the fact that human surroundings contain much more than the

---

thinking about how the objects that one takes to exist in one's surroundings might not exist in the 'surroundings in-themselves' then one is having thoughts which are not 'typical thoughts'; it is possible that such thoughts could lead to an accurate *general conception.*

normal human perceptual apparatus can perceive. This realisation, in itself, is all that we need to take us to the conclusion that a human's conception of the nature of their surroundings is highly likely to be misguided. For, most humans naturally conceive of their surroundings to be as they appear to them to be; they *do not* naturally conceive of their surroundings to be containing much more than appears to be there.[8]

3.1.2 Inevitable Constraint 1: Interiority

The first way in which human perceptual connections are inevitably constrained is their inability to access interiority. Imagine that one is standing at the side of a road and watching tens of thousands of humans running a marathon; one is unable to perceptually connect to the interiority states of these humans. Presumably these humans contain interiority states – states such as the feeling of pain, the feeling of aching and the feeling of pins and needles – but one lacks a perceptual connection which would enable one to connect to these states.

One does have perceptual connections which enable one to perceive that a human is almost certainly suffering from excruciating pain – connections to a broken leg, streams of blood,

---

[8] You might find this to be slightly contradictory. On the one hand, humans have gained the knowledge that non-human animals can hear sounds that humans are unable to hear; on the other hand, humans typically do not conceive of their surroundings to be containing much more than appears to be there. How can this be? Recall that we live in an epoch in which there is a violent clash between human knowledge of their surroundings and human interactions with their surroundings; an epoch in which knowledge has been gained but has yet to have a widespread effect on the general conception that humans have of their surroundings. This apparent contradiction is an aspect of the violent clash.

screaming – but this is a very different thing from having a perceptual connection to the *feeling* of pain. It seems obviously true that states of raw feeling like this are *in principle* imperceptible to human perceptual connections. Indeed, this is the conclusion that we reached in *Section 2.5*. So, the inability to access interiority means that the human perceptual apparatus is inevitably constrained.[9]

3.1.3 Inevitable Constraint 2: Limited Scope

The second way in which human perceptual connections are inevitably constrained is that a perceptual connection is, by its very nature, limited in the scope of the connections it can make. However much one technologically modifies these connections it is inevitable that there will be parts of the surroundings which are outside the scope of the connection.

In order to appreciate the limited scope of human perceptual connections it is useful to adopt the evolutionary perspective. One could reasonably believe that human perceptual connections have evolved in order to perceive the kinds of things that are of survival interest. The belief that the human perceptual apparatus has been moulded by evolution to perceive the universe from a narrow perspective is the central claim of the evolutionary epistemology paradigm. The evolutionary epistemologist claims that we have only developed 'organs' for

---

[9] In *Section 5.3.4* I defend the idea that humans can become aware of feeling states which are located outside of their body. This position is consistent with the interiority inevitable constraint because it entails that humans are unable to know where in the universe the feeling states are located; 'become directly aware of' and having a 'perceptual connection' are two different things.

those aspects of our surroundings which it was imperative for our ancestors to take account of if they were to survive. For example, G. G. Simpson (cited in Sjolander, 1997, p.596) claims that: "The monkey that had no realistic perception of the branch he was jumping for was soon a dead monkey – and did not belong to our ancestors." Whilst, Konrad Lorenz (1977, p.7) claims that: "our cognitive apparatus resembles that of a primitive hunter of whales or seals, who only knows about his quarry what it is of practical value for his purposes to know." The evolutionary epistemologist believes that human surroundings exist, that humans don't have access to all of these surroundings, but that what they have access to actually exist. Thus, Lorenz (1997, p.7) continues:

> Yet what little our sense organs and nervous system have permitted us to learn has proved its value over endless years of experience, and *we may trust it* – as far as it goes. For we must assume that reality also has many other aspects which are not vital for us, barbaric seal hunters that we are, to know, and for which we have no 'organ', because we have not been compelled in the course of our evolution to develop means of adapting to them.

In this way evolutionary epistemology is both more modest and more ambitious than Kantian transcendental idealism. According to Gerhard Vollmer (1984, p.81) evolutionary epistemology, "is *more modest* in not claiming necessary truth or objectivity… [and] *more pretentious*… [as] we hope to find, at least, some *truth* about the world as it is, not only as it appears to us." These claims are justified by making the distinction between experiential knowledge and scientific knowledge. Human experiential knowledge is wholly dependent on inherited cognitive structures and thus gives rise to a species 'mesocosm'. Vollmer (1984, p.87) states that, "Every organism has its own cognitive niche or ambient, and so does man...Our mesocosm is that section of the real world we cope with in perceiving and acting, sensually and motorially... [it is] a world of medium dimensions." It is the presence of the mesocosm which means that objective truth claims cannot be made on the basis of experiential knowledge; this knowledge is acquired in an uncritical way.

However, according to this view scientific knowledge is abstract and critical; it is not genetically determined in the manner of human experiential knowledge. As Vollmer (1904, p.81) puts it: "It would not make sense to ask for the biological roots of relativity theory, of quantum chromodynamics, or of any other theory of modern science." The pretentious claim of evolutionary epistemology is that we can use reason and science to transcend our 'mesocosm' and gain insight into the 'thing-in-itself' – true knowledge about the universe as it actually is that is barred to the transcendental idealist, who believes that the mesocosm is an impenetrable barrier. Nevertheless, as Lorenz (1982, p.139) puts it: we are still "lacking the receptive organs for infinitely much that is actual."

Another variant of the limited scope inevitable constraint is the idea that human perceptual connections inevitably perceive their surroundings as containing both secondary qualities and primary qualities, when in reality the 'surroundings in-themselves' only contain primary qualities. If this is so, then the perceptual connections are limited in that they entail an act of transformation; the process of connecting to primary qualities involves a transformation which is imparted by the connection itself – the connection transforms one's perceived surroundings by endowing them with secondary qualities. If humans inevitably perceive their surroundings as coloured, when the 'surroundings in-themselves' are wholly uncoloured, then the human perceptual apparatus is inevitably constrained.

3.1.4 Inevitable Constraint 3: Temporal

The human perceptual apparatus is also inevitably constrained because it has in-built temporal constraints; it is only able to perceive movements from an exceptionally narrow temporal perspective. This is possibly a hard thing to envision; how is one to get a handle on this inevitable constraint? Let us start with the evolutionary perspective.

The universe has been evolving and moving in various ways for billions of years; in contrast, the average contemporary human will be lucky to reach the age of one hundred years. One can barely comprehend what it would be like to perceive a movement that spanned a thousand years or a million years – in contrast to the movements which our perceptual apparatus has evolved to perceive, such as *the running of a wild animal towards one,* which is a movement which lasts a matter of seconds – but such long-term movements clearly exist. The human perceptual apparatus has evolved to connect to short-term movements and is unable to connect to long-term movements.

How exactly does the exceptionally narrow temporal perspective from which humans are able to perceive movements in their surroundings inevitably constrain their perceptions? It is perhaps helpful to start by considering a relatively short-term movement which a human could, *in principle*, be able to perceive. So, there is no reason why one could not perceive the movement pattern that is the Earth taking 365 days to move around the Sun. If one was located in an appropriately positioned space station, and was able to continuously observe for 365 days, then one would be able to perceive this movement (of course nearly-all, if not all, humans alive at the moment would not be able to do this as they need to sleep roughly every 24 hours). Of course, when we start to consider slightly longer-term movement patterns, those that exceed the lifespan of a human, then it is obviously the case that it is impossible for a human to be able to observe these movement patterns (a human can only observe whilst they are alive!).

Why is this important? If a human only has perceptual access to a small temporal slice of a movement, then that human is not in a position to accurately judge the nature of the movement. The inability of humans to access long-term movements means that they are likely to conceive of much of their surroundings as mechanistic – this is because the small segments of movements that humans are able to access *appear to them* to be mechanistic. If humans had perceptual access over a longer temporal window then all of the movements which humans perceive in their surroundings might appear to them to be non-mechanistic. So, the inevitable temporal limits of the human perceptual apparatus can easily lead one to conceptualise the vast majority of one's surroundings as mechanistic – as very different from humans.

I will use an example to clarify this point. Let us consider a series of very short-term movements such as all the movements of the players on the pitch in a 90 minute football match. If you perceived this series of movements over a 90 minute period you would, no doubt, conclude that they were non-mechanistic. However, if you only had perceptual access to the first second of the match what would you conclude? The movements which you were able to perceive within this temporal window would not be of long enough duration to enable you to conceive of them to be non-mechanistic. You would, no doubt, conclude that the movements were mechanistic. It is only if you had a longer time slice of perceptual data that you would be able to conclude that the movement which you previously conceived as mechanistic is actually part of a much longer duration movement pattern which you would now wish to assert is non-mechanistic.

I hope you can see this. The universe is the 90 minute football match. All of the perceptions that a human can have of the universe occupy the first second of the match. Humans form their conceptions of the universe based on this first second. But the universe isn't the first second – the universe is the whole 90 minute match! The human perceptual apparatus is clearly inevitably temporally constrained.

3.2   The connection to the 'surroundings in-themselves'

In the previous sections I have assumed that human perceptual connections connect to the 'surroundings in-themselves'. I have contended that even if this is so that human perceptual connections are inevitably constrained in three different ways. In this section I will be defending this assumption. However, I will also be defending the idea that the objects that humans typically take to exist in their surroundings – objects such as trees, pencils, frogs, other humans, cats, planets and helicopters – do not exist in the 'surroundings in-themselves'.

What exactly does it mean for perceptual connections to connect to the 'surroundings in-themselves'? This question is typically framed in the following way. Human perceptual connections reveal certain objects to exist in their surroundings – dogs, trees, chairs, helicopters, other humans, planets, screams, etc. If these objects exist in the 'surroundings in-themselves' – exist when the perceptual connections do not exist – then human perceptual connections connect to the 'surroundings in-themselves' (as we have seen, even if perceptual connections connect to the 'surroundings in-themselves' they are still inevitably constrained). This view is often referred to as direct realism.

The alternative is to believe that human perceptual connections fail to connect to the 'surroundings in-themselves'. According to this view, when the perceptual connections do not exist the objects do not exist. This view is typically referred to as 'conceptualism in perception' and is a variety of anti-realism. There are a couple of rather fruity versions of this position. According to Kantian transcendental idealism the 'surroundings in-themselves' do not contain dogs, trees, and screams – they are aspatial and atemporal; it is the perceptual connection which brings space, time and objects into existence. According to the wavefunction collapse theory the 'surroundings in-themselves' do not contain dogs, trees, and screams – but they do contain spatio-temporal wavefunctions. According to this view perceptual connections collapse wavefunctions into determinate form and thereby bring objects into existence.

As I've said this is the typical framing of the issue. Perceptual connections connect to the 'surroundings in-themselves' and the objects exist whether they are perceived or not. Or, perceptual connections do not connect to the 'surroundings in-themselves' and the objects only exist when the perceptual connection exists. I will be defending another possibility – the idea that the objects that human perceptual connections connect to *do not* exist in the 'surroundings in-themselves', but that human perceptual connections *do* connect to the 'surroundings in-themselves'.

In agreement with the realist I believe that the 'surroundings in-themselves' are spatial, temporal and continuously collapsed; in agreement with the anti-realist I believe that the 'surroundings in-themselves' are objectless.[10] The realist believes that perceptual connections reveal pre-existing objects, and the anti-realist believes that perceptual connections create objects; in this respect I disagree with both the realist and the anti-realist – the 'surroundings in-themselves' are objectless *and* perceptual connections are wholly devoid of concepts/objects. If this is so, how do the objects get into the picture? The objects come into existence when the perceptual connections are rationalised, when they are thought about. This scenario might be a hard thing to envisit. How can one's perceptual connections connect to their surroundings without connecting to objects – objects such as dogs, frogs, books and trees? Let me try and explain.

Imagine that the totality of one's surroundings (the 'surroundings in-themselves') is a 'blobject' – an enormous collection of spatio-temporally located particles. In the 'blobject' there are no distinct objects; there are just particles which are arranged in lots of different ways. Imagine one particular part of

---

[10] By 'objectless' I mean that the 'surroundings in-themselves' do not contain dogs, trees, humans and screams. On this view, one can still refer to the totality of the 'surroundings in-themselves' as a single object – as a 'blobject'.

the 'blobject' – this part will contain a particular arrangement of particles. This arrangement will be very similar to the arrangements which exist in some of the other parts of the 'blobject', and very dissimilar to other arrangements.

Imagine that the 'blobject' is the 'surroundings in-themselves'. When a perceptual connection is forged between a perceptual apparatus and the 'blobject' what does the perceptual apparatus connect to? One can very plausibly believe that it does not connect to a particular object – an object such as a dog, a tree, or a scream. On the contrary, one can believe that it connects to a particular region of the 'blobject' itself – a region which is objectless and boundary-less. The only boundary would be the scope of the connection enabled by the perceptual apparatus – it can only illuminate a tiny segment of the 'blobject', it cannot penetrate into its 'interior', and it can only perceive it from a very narrow temporal perspective. Let us consider a particular visual perceptual connection; this connection involves a perceptual apparatus illuminating a particular arrangement of particles/colours. This connection entails perceiving the 'blobject' as containing yellow in a particular location, green slightly to the left and slightly closer, blue to the left of the green, behind these colours there is a lot of brown, above the brown there is a thin layer of white, above the white there is a thin layer of black, above the black there is a large area of blue which extends away into the distance, and above this blue there is some black. Let us assume that this is all that exists in the visual perceptual connection (a typical human visual perception connection would need a much more detailed description than this!).

I hope that this example is sufficient for you to see the point. The visual perceptual connection connects to particles/colours which are arranged in a particular way, but not to particular objects. These perceptual connections are objectless and boundary-less; they do not involve thoughts or concepts.[11] It is when these connections feed into a process of thought that concepts and objects arise. One thinks something like: that arrangement of white and black is a book, that arrangement of green and yellow is a shirt, that arrangement of brown and white is a bear, that arrangement of black and white is a cat, that yellow circular arrangement is the Sun, etc. There are no objects either in the 'blobject' or in the perceptual connections. This sharp disparity between perceptual connections and concepts is argued for by E. J. Lowe (2000, p.182) who states that:

> pigeons can be trained to discriminate visually between triangles and squares, but it would be extravagant to suggest that they therefore possess the concepts of trian-gularity and squareness. For example, we should not attribute possession of the concept of a *tree* to someone unless we are prepared to attribute to that person certain general beliefs concerning trees, such as that trees are living things which grow from the ground and have branches, roots and leaves. The mere ability to discrimi-nate visually between trees and other objects, such as rocks, and to engage in distinctive behaviour with respect

---

[11] Is it even intelligible to believe that a connection between a perceptual apparatus and its surroundings involves thoughts? It is very hard to make any sense of how this could be so. Is it intelligible to believe that such a connection involves concepts? This seems to depend on how one conceptualises a concept, and on how one conceptualises the connection between a perceptual apparatus and the 'surroundings in-themselves'.

to them, such as nest-building, is not enough to constitute possession of the concept of a tree.

According to the view I am defending, when humans
perceive the 'blobject' they perceive it in a particular way – their
perceptual apparatus provides *the possibility* for a *certain range
of objects* to come into existence, which will occur if the perceptions are thought about. Similarly, when pigeons perceive the
'blobject' their perceptual apparatus provides *a different range
of possibilities* for objects to come into existence if these
perceptions are thought about. Pigeons can be trained to
identify the parts of the 'blobject' *which humans call* 'triangles'
and 'squares', but their perceptions can be very different to
human perceptions, and there is no reason to expect that
pigeons bring the objects 'triangle' and 'square' into existence
when they think about their perceptions of the part of the
'blobject' *that humans call* 'triangle' and 'square'.

According to this view what a perceptual apparatus
connects to exists for that perceptual apparatus (there cannot be
inaccurate perception).[12] As a particular part of the 'blobject' can
be simultaneously perceived very differently by diverse sets of
perceptual apparatus this means that there are not more
accurate or less accurate perceptions; there are just different
perceptions. Whatever is connected to in a particular moment is

---

[12] Following a particular perception one might think 'there is an oasis in front of
me'. On investigating one might find that there is no oasis. This means that the
thought was inaccurate; however, the perceptual connection that initiated the
thought existed (it wasn't inaccurate). One simply thought that this connection
revealed a particular arrangement of the 'blobject' (one's 'oasis' concept) when
in actuality it did not.

a state of the 'blobject' that exists in that particular moment for the connecting perceptual apparatus. Given that humans have similar brains and similar senses one would expect that all typical humans will have a perceptual apparatus that *provides the possibility for the same range of objects* to come into existence. And if humans live in a similar culture/social setting then one would expect that these humans would conceptualise their perceptions into similar objects (for example: books, ships, cats, paper clips). However, if two humans come from a radically different culture/social setting then one would expect that when they perceived the same area of the 'blobject' that they would conceptualise it into different sets of objects. For example, according to one widely discussed case, when the European sailing ships arrived in the New World the natives on the beaches initially couldn't see the ships as they were approaching the shore. They couldn't see the ships because they had no conception of what such an object could be; to them it was simply part of the 'blobject' – not an object – not a 'ship'.

I have outlined the view that human perceptual connections connect to the 'surroundings in-themselves', that the 'surroundings in-themselves' are the 'blobject', that perception does not entail objects or concepts, and that objects are concepts which require a process of thought for their existence. I will now address a particular worry which you might have with the view that I have outlined – the 'veridicality worry' – and then I will try and convince you to believe in the 'blobject'.

Why might one be tempted to believe that there are objects in a universe which has no perceivers? This temptation arises from the 'veridicality worry':

When you say you see a cat and I say I see a badger (speaking of the same area of space) don't we want to be able to say one of us is wrong and (possibly) one of us right?

There is a good chance that you might believe that the answer to this question is 'yes'. After all, you probably went to school at a young age and were shown pictures of cats and badgers and were told '*this is a cat, and this is a badger*'. One might also have grown up with a pet cat and one's parents told one '*this is a cat, it is not a badger – if you see the cat stroke it, if you see a badger run*'. One probably wasn't told at school, or by one's parents, that the universe is a 'blobject' and that there are no objects without perceivers; such deep issues are a bit much for schoolchildren. In short, humans are typically brought up to believe that there is a way *that things are* in their surroundings – that if Tim sees a cat where Jane sees a badger that one of them must be wrong.

It is important to realise that there is an important difference between there *not being a fact of the matter* as to whether a particular part of the 'blobject' contains a cat or a badger, and the *likelihood* of Tim seeing a cat where Jane sees a badger. As has already been discussed, if we assume that Tim and Jane are humans with a typical human visual perceptual apparatus, who have grown up in the same culture, then, bar some exceedingly unlikely set of circumstances, where Tim sees a cat Jane will see a cat, and where Tim sees a badger Jane will see a badger. Yet, there is no cat or badger in the 'surroundings in-themselves'.[13]

---

[13] In other words, *if* Tim and Jane have a sufficiently similar perceptual apparatus *and* they evolved in a sufficiently similar culture, then there will be a fact of the

One might still have the 'veridicality worry'. One might want to know what exactly *it is* that exists where Tim and Jane see a cat, if it is not a cat. Let me be clear, 'a cat', and all of the other objects that humans refer to – objects such as tables, pens, humans, helicopters, yurts, rice, pigs, badgers, and trees – are parts of the 'blobject' that humans conceptualise as objects. These objects are parts of the observable part of human surroundings. In the area of the 'blobject' where a human perceives 'a cat' there will be a particular arrangement of particles and emptiness which can be seen as 'a cat' by many humans. In this area of the 'blobject' there will also be states which are unobservable – 'feeling states', 'awareness', 'thoughts'.[14] In other words, there is a matter of fact as to whether a particular part of the 'blobject' is thinking, is aware, and contains particular 'feeling states'. Where a human perceives 'a cat' there is no cat, but there can be an area of thought, an area of awareness, and particular 'feeling states'.[15] If one has the 'veridicality worry' the realisation that this is so can hopefully help one to overcome it.

Now let us consider why one should believe that the 'blobject' is the most plausible description of the relationship between humans and their surroundings. Firstly, let us consider all of the different perceivers that might simultaneously perceive

---

matter as to whether a particular part of the 'blobject' contains a badger or a cat *for Tim and Jane*. But there is no fact of the matter as to whether a badger or a cat exists in the 'blobject' itself.

[14] If one is not convinced that the part of the 'blobject' that is 'a cat' has all of these states then one can simply replace 'a cat' with 'a human'.

[15] In *Section 5.4.6* I will also claim that a particular part of the 'blobject' contains a multitude of colours and sounds.

the same part of the universe – humans, dolphins, bats, dogs, Martians, birds, naked mole rats, frogs, chipmunks, cats, hedgehogs, lions, pigs, and so on. The most plausible thing to believe is that all of these different perceivers take this part to contain very different objects. How can we make sense of this? We can believe that the 'surroundings in-themselves' do not contain distinct objects; they just contain a particular arrangement of particles and emptiness which can be perceptually connected to in a multitude of diverse ways. In *Section 2.3* we saw how this can be so – in the absence of organisms there are no environments. It is biology that determines which arrangements of particles and emptiness are connected to.

The evolutionary perspective also provides a good reason to believe in the 'blobject'. Imagine the moment of the Big Bang; imagine the evolutionary processes which have been going on since this event – the processes which have created what we call solar systems and planets. Let us assume that in the immediate aftermath of the Big Bang that there were no perceivers.[16] What was the universe like at this time? Were there objects? Or were there just particles arranged in a particular way at a particular moment in time (a 'blobject')?

Of course, humans look back in time and say things like: *the planet Earth must have existed before perceivers existed on the Earth so the object Earth existed before there were perceiv-*

---

[16] It is worth noting that in *Chapter Six* I defend the idea of panperceptualism. However, it is also worth noting that panperceptualism doesn't entail the existence of objects (objects require thoughts).

ers.[17] From our discussion of the 'blobject' one will realise that this just won't do. Before there were perceivers there are very good reasons to believe that what we call the Earth wasn't a distinct object, it was just a part of the 'blobject'. It is easy to think of the Earth as an object which existed before there were perceivers, just as it is easy to think that the Earth is flat, or that the Earth is the centre around which the Sun revolves. Why is this easy to believe? It is easy to believe because the Earth seems to be a solid object which is surrounded by emptiness, emptiness which exists until another solid object exists – an object we call the moon.

Why do we need to rid of ourselves of this way of thinking? Our scientific investigations into our surroundings have revealed that they are almost entirely empty space. An atom is a tiny nucleus surrounded by distant electrons which circle the nucleus; atoms are overwhelmingly emptiness. So, to believe that the Earth is an object because it is surrounded by emptiness does not stand up to scrutiny; for, the entire universe is composed of tiny particles all of which are surrounded by vast swathes of emptiness. Before there were perceivers the universe was just particles and emptiness. The most accurate description of the evolution of the universe after the Big Bang is: *the continuous bringing forth of a different arrangement of particles and emptiness.*

The state of the universe before there were perceivers is the 'surroundings in-themselves'. The 'surroundings in-themselves' did not magically transform themselves when perceivers came

---

[17] Recall that I am defending the view that perceivers are necessary but not sufficient for the existence of objects. This means that the word 'perceivers' in this sentence could be replaced with 'thinkers'.

into existence; the universe today is still a 'blobject' – at any moment in time it is a particular arrangement of particles and emptiness. It is just that in the universe today perceptions of the universe exist which enter the realm of thought and thereby bring objects into existence. This 'bringing of objects into existence' might sound a bit mysterious, but all it means is that there are particular arrangements of the universe which believe that the universe is comprised of distinct objects. In effect, these thoughts can be thought of as 'carving' the 'blobject' into small segments – they conceptualise the universe as lots of individual objects, when the reality is that in-itself the universe is a singular 'blobject' which can be carved up in a multitude of ways.

I hope to have shown that one can simultaneously have the following two beliefs: the human perceptual apparatus can connect to 'surroundings in-themselves' and the 'surroundings in-themselves' are wholly devoid of objects. Furthermore, I hope to have shown that there are very good reasons why one should have these two beliefs.

3.3   *The general conception*

The general conception that a human has of their surroundings is a particular set of beliefs concerning those surroundings; a general conception can also be thought of as a world view. There are a plethora of world views, some are radically different and some are only subtlety different. In a given epoch it is likely that a particular world view will be dominant; that is to say, it is likely that a particular world view will be instantiated in the majority of living humans.

In order to elucidate the relationship between a human's perceptual connections with their surroundings and their general conception of those surroundings I will briefly outline two particular world views – the Primal World View and the Modern World View. It is widely accepted that in the past the Primal World View was dominant and that the Modern World View replaced the Primal World View. Richard Tarnas (2007, p.16) describes these two views:

> The primal human perceives the surrounding world as permeated with meaning...Spirits are seen in the forest, presences are felt in the wind and the ocean, the river, the mountain...The primal world is ensouled. It communicates and has purposes. It is pregnant with signs and symbols, implications and intentions. The world is animated by the same psychologically resonant realities that human beings experience within themselves...Creative and responsive intelligence, spirit and soul, meaning and purpose are everywhere.

> The modern mind experiences a fundamental division between a subjective human self and an objective external world...Whatever beauty and value that human beings may perceive in the universe, that universe is in itself mere matter in motion, mechanistic and purposeless, ruled by chance and necessity...The world outside the human being lacks conscious intelligence, it lacks interiority, and it lacks intrinsic meaning and purpose.

In the Primal World View a human's perceptual connections with their surroundings are conceptualised in a way which leads to the belief that those surroundings are very similar to humans.[18] In contrast to this, in the Modern World View a human's perceptual connections with their surroundings are conceptualised in a way which leads to the belief that the vast majority of those surroundings are very dissimilar to humans.[19] This means that human perceptual connections can be conceptualised very differently. Whilst all typical biological humans have very similar perceptual connections (determined by human physiology), humans can both carve these perceptions up into different objects (determined by their cultural setting), and conceive of the surroundings which are revealed very differently. Humans can conceive of their surroundings as very similar to humans, or as radically different to humans. We currently live in a 'scientific' epoch in which humans typically conceive of their surroundings as very un-human like.

It should be noted that whilst the Modern World View supplanted the Primal World View as the dominant worldview, the Primal World View still exists. Furthermore, the fact that a particular world view has supplanted a rival view for dominance obviously does not entail that the supplanting view is the best guide to the actual nature of human surroundings. Indeed, as we saw in *Chapter One,* it is possible that humans have a special place in the universe because they do not consider themselves to

---

[18] In *Chapter One* I discussed the question of *what it means to be human.* It seems that in the distant past, when only the Primal World View existed, that whilst there were 'biological humans' there weren't any humans.

[19] One could have the Modern World View and still believe that some non-human animals are similar to humans.

be natural. Now, we can see that the move from the Primal World View to the Modern World View is a move of potentially important significance. It seems to be a move that is associated with the transition from humans considering themselves to be natural to not considering themselves to be natural.[20] In *Chapter Eight,* when we consider the philosophy of Friedrich Hölderlin, we will see that the Modern World View can be seen as a temporary disharmony; according to this view, in the future the Primal World View will return to dominance.

3.4 Conception and perception

In *Section 3.1* we focused on a particular aspect of the conception that a human has of their surroundings – the way that the perceptual connections that exist between a human and their surroundings are severely inevitably constrained. In *Section 3.2* we explored the resulting belief (when these perceptions are conceptualised) that the universe contains objects such as trees, frogs and other humans; I contended that the 'surroundings in-themselves' are wholly devoid of objects. In *Section 3.3* we looked at the *general* conception that a human has of their surroundings. The purpose of this section is to explore the implications of all of this.

The general conception that a human has of their surroundings is parasitic on their perceptual connections; we concluded that these connections are severely inevitably constrained. This means that the general conception that a

---

[20] In *Section 7.1* I will further explore and defend the claim that the first humans to evolve considered themselves to be natural, whereas today we live in an epoch in which the overwhelming majority of humans consider themselves to be not natural.

human has of their surroundings is *unlikely* to be a good guide to the nature of the 'surroundings in-themselves' (even though I have suggested that the connections connect to the 'surroundings in-themselves').

When one realises that one's perceptual apparatus is inevitably constrained one can change one's *general conception* in order to try and offset this constraint. From the evolutionary perspective we have a very good reason to believe that *in reality* humans are actually quite similar to their surroundings. However, the inevitable constraints which we have identified cause humans *to conceive of* their surroundings as dissimilar to themselves. So, it is likely that when one realises that the inevitable constraints exist that one will change one's *general conception* so that one starts to believe that one's surroundings are more similar to one.[21] So, when one realises that the 'interiority' constraint exists, one can believe that one's surroundings are also pervaded with interiority; when one realises that the 'temporal' constraint exists, one can believe that one's surroundings are wholly non-mechanistic; and when one realises that the 'limited scope' constraint exists this can ultimately lead one to believe that one is part of a single object (a 'blobject').

We can now clearly see why there will be many different world views. This is because identical perceptions can be conceptualised very differently. It is also because many humans do not realise that the human perceptual apparatus is inevitably constrained; so these humans are likely to believe that human surroundings are radically different to humans. Whilst those humans that realise that the inevitable constraint exists have a

---

[21] If one rejects the evolutionary perspective (or is unable to fully comprehend it) then this might not be so likely.

range of different ways in which they can adjust their beliefs in an attempt to compensate for its effects. So, it is not surprising that there are a variety of contemporary views concerning how similar humans are to their surroundings; in *Section 3.5* we will consider several of these views.

3.5 Various views concerning how similar humans are to their surroundings

The considerations from the previous sections have put us in a good position to understand why there are a variety of views concerning how similar humans are to their surroundings. The purpose of this section is to explore several of these world views.

3.5.1 Physicalism

A widespread contemporary world view concerning human surroundings (and humans themselves) is 'physicalism' – also known as 'materialism'. This view is typically closely associated with the Modern World View, rather than with the Primal World View. This association is perhaps surprising, for physicalism is the view that *humans and all of their surroundings are made of a single type of stuff* – the 'physical'; given this one might think that the physicalist would conceive of their surroundings as very similar to them (as in the Primal World View), rather than as radically different to them (as in the Modern World View). On an *alternative* view, where not *everything* that exists is physical – where there are some physical things and some non-physical things – then it is much easier to make sense of radical differences in the universe, such as the hypothesised radical differences between humans and their surroundings.

The curious association between physicalism and the Modern World View raises the following question: What does it mean for a thing to be physical? According to the *theory-based conception* (Stoljar, 2009): "A property is physical if and only if it either is the sort of property that physical theory tells us about or else is a property which metaphysically (or logically) supervenes on the sort of property that physical theory tells us about". According to the *object-based conception* (Stoljar, 2009): "A property is physical if and only if it either is the sort of property required by a complete account of the intrinsic nature of paradigmatic physical objects and their constituents or else is a property which metaphysically (or logically) supervenes on the sort of property required by a complete account of the intrinsic nature of paradigmatic physical objects and their constituents".

The *theory-based conception* of the physical faces an obvious problem. For, according to Hempel's dilemma, this conception leads to physicalism either being trivially true or false. If physicalism is defined in terms of contemporary (incomplete) physics it will be false. However, if physicalism is defined in terms of a completed physics then the concept is trivial as nobody knows which properties would be present in a completed physics. A 'complete' physics could include properties such as 'what-it-is-likeness' properties, 'awareness' properties, and other properties which we currently cannot even conceptualise. Furthermore, a legitimate worry seems to be whether humans could possibly ever know when they have attained a 'completed physics'. It seems possible that in the future humans could think they have attained a 'completed physics' only to later realise that such a theory was actually incomplete.

For this reason the *object-based conception* of the physical is the more plausible option. According to this view a property is physical if a complete account of paradigmatic physical objects includes that property. Such a complete account clearly exists, the question is: Can a complete account of paradigmatic physical objects be known by a human? If one realises that the way in which humans gain information about their surroundings is inevitably constrained by their perceptual apparatus then one will have a very compelling reason to doubt that such a complete account can be attained.

As we saw in *Section 2.5,* many philosophers such as Kant, Strawson, Unger and Russell, have realised that the human perceptual apparatus is inevitably constrained and therefore that the nature of human surroundings is forever beyond the bounds of human knowledge. In *Section 3.1* we also explored a range of inevitable constraints which limit the ability of humans to gain knowledge about the nature of their surroundings. So, the *object-based conception* of the physical is seemingly just as inadequate as the *theory-based conception* of the physical.

The realisation of the extent to which a human's knowledge of the nature of their surroundings is limited opens up various possibilities. One could obviously continue to assert that physicalism is true – that the universe only contains physical properties – it is just that there are physical properties which humans are perceptually barred from. This acceptance of *unknown physical properties* clearly opens the floodgates to a number of varieties of physicalism. The varieties which I will be considering are 'emergent physicalism', 'panwhat-it-is-likeness physicalism', and 'panawareness physicalism'. Alternatively, one could reject the idea of monism and advocate dualism. In the following sections these various possibilities will be considered.

3.5.2 Emergent physicalism

It is possible to believe that the entire universe is wholly comprised of physical objects, that paradigmatic physical objects such as trees and tables might have physical properties which humans are perceptually barred from knowing anything about, and to believe that these unknown physical properties (if they exist) are wholly devoid of 'what-it-is-likeness', 'awareness', and any other type of 'mentality' or 'interiority'.

According to this view humans are physical objects, and humans have 'what-it-is-likeness', 'awareness', 'mentality' and 'interiority'; this means that some physical objects do have these attributes. This view – that the physical is mostly devoid of 'interiority' (I will use this term as a shorthand for all of the various phenomena listed in the previous sentence) and that it is only in certain parts of the physical that 'interiority' emerges – we can refer to as 'emergent physicalism'.

This view seems to be intuitively plausible as one probably ordinarily does not suppose that objects such as tables contain 'interiority'. However, the emergent physicalist needs to give a plausible account of the coherence of such a disjointed world view. How is it possible that physical stuff which itself is wholly devoid of 'interiority' when it is arranged in a certain way can generate 'interiority' when it is arranged slightly differently? Such emergence is not beyond the realms of possibility, but giving a convincing account of the intelligibility of such emergence is deeply problematic; indeed, some philosophers go so far as to argue that such emergence is incoherent. For example, Galen Strawson (2006, p.12) writes:

Does this conception of emergence make sense? I think that it is very, very hard to understand what it is supposed to involve. I think that it is incoherent, in fact, and that this general way of talking of emergence has acquired an air of plausibility (or at least possibility) for some simply because it has been appealed to many times in the face of a seeming mystery.

It is common for certain properties to emerge with complexity; this is unproblematic. The question is about the intelligibility of the notion of 'interiority' properties emerging out of arrangements of stuff which are wholly devoid of 'interiority'. Strawson (2006, pp.13-15) claims that:

liquidity is a truly emergent property of certain groups of H_2O molecules. It is not there at the bottom of things, and then it is there... But can we hope to understand the alleged emergence of experiential phenomena from non-experiential phenomena by reference to such models? I don't think so. The emergent character of liquidity relative to its non-liquid constituents does indeed seem shiningly easy to grasp...But when we return to the case of experience, and look for an analogy of the right size of momentousness, as it were, it seems that we can't make do with things like liquidity, where we move within a completely conceptually homogenous (non-heterogeneous) set of notions. We need an analogy on a wholly different scale if we are to get any imaginative grip on the supposed move from the non-experiential to the experiential.

Strawson (2006, p.15) claims that an analogy of an appropriate scale would be the emergence of the extended out of an arrangement of the wholly unextended, and asserts that this "should be rejected as absurd". If this is an analogy of the appropriate scale then the emergence of 'interiority' does seem to be absurd; the question is whether it is an analogy of the appropriate scale. Catherine Wilson (2006, p.182) asserts that: "it is hard to see why it is *impossible* that what one needs for there to be experiences in the universe is a brain made of insentient molecules put together in a certain way". One should agree with Wilson that one should not assert that the emergence of 'interiority' out of that which is wholly devoid of 'interiority' is impossible; given the constraints on human knowledge, there are very few things which humans can confidently claim to be impossible. The real issue seems to be not impossibility but likelihood; to say that something is 'not impossible' is not to say that there are good reasons for believing it.

The idea that 'interiority' can emerge out of that which is wholly devoid of 'interiority' is clearly a contentious one. All that the supporter of such emergence can do is assert that they have faith that it happens, and cite as analogies less contentious cases of emergence. Whilst, from the other side, accounts of emergence such as liquidity are not going to convince those who think that such analogies are misplaced, and that the emergence of 'interiority' is incoherent.

Given that the idea that 'interiority' emerges is deeply problematic it is helpful to consider why the emergent physicalist believes that such emergence occurs. In other words, the question is: Why does the emergent physicalist *deny* that all of the physical has 'interiority'? The main motivation for this denial seems to be that one is the *only* part of the universe that one indubitably knows has 'interiority'; starting with themselves the emergent physicalist then rationalises that parts of their surroundings which resemble them probably also have 'interiority' – parts such as other humans and some non-human animals. In contrast, a table and a tree seem to be most unlike them, so they conclude that such things don't contain 'interiority'.

The problem with this line of reasoning by analogy is that a human's perceptions of their surroundings are inevitably constrained by their perceptual apparatus; if their perceptions were not so constrained then a human might conclude that all of their surroundings resembled them, and hence that all of their surroundings contained 'interiority'. In short, the way that the human perceptual apparatus works causes humans to conceive of some of their surroundings as similar to themselves and other parts of their surroundings as dissimilar to themselves, and the emergent physicalist is only happy to suppose that parts of their surroundings which appear to be sufficiently similar to themselves contain interiority. When the inevitable constraint is acknowledged there is a good reason for believing that the suppositions of the emergent physicalist don't match a real division within the universe.

3.5.3 *Panwhat-it-is-likeness physicalism*

In the previous section we saw that that the emergent physicalist finds no good reason to analogically extend states of 'interiority' to all parts of the physical. In contrast, the panwhat-it-is-likeness physicalist is quite happy to extend the scope of interiority to encompass all of the physical. This is intellectually appealing because it means that one does not need to believe that that which is wholly devoid of interiority when arranged in a particular way can give rise to interiority when it is arranged slightly differently. The panwhat-it-is-likeness physicalist believes that the emergent physicalist has gone wrong because they fail to acknowledge the inevitable perceptual constraint; if they did then they could overcome the problem they have created – the 'emergence of interiority' problem – by extending 'interiority' from *some* of the physical to *all* of the physical.

In *Section 3.5.2* I stated that I was using the term 'interiority' as a shorthand for various phenomena – 'what-it-is-likeness', 'awareness', or any other type of 'mentality' or 'interiority'. The panwhat-it-is-likeness physicalist believes that the physical is pervaded with 'what-it-is-likeness', and 'what-it-is-likeness' is a particular type of interiority. However, the panwhat-it-is-likeness physicalist denies that the physical is pervaded with 'awareness'.[22]

What exactly does it mean to believe that all of the physical is pervaded by 'what-it-is-likeness'? What it means in a broad sense is that every part of the physical contains qualitative feeling – states analogous to those that a human might refer to as the feeling of pain, the feeling of tingling, and the feeling of 'pins and needles'. Of course, in denying that awareness pervades the physical the panwhat-it-is-likeness physicalist needs to give an account of the emergence of awareness out of the wholly unaware. However, the panwhat-it-is-likeness physicalist does not face the much bigger problem which is faced by the emergent physicalist – that of giving an account of the emergence of interiority out of that which is wholly devoid of interiority.

---

[22] The difference between 'what-it-is-likeness', 'awareness' and 'mentality' is explored in *Chapter Four*.

3.5.4 *Panawareness physicalism*

In the previous section we saw that the panwhat-it-is-likeness physicalist extends the states of what-it-is-likeness that they know exist within themselves to all parts of the physical. This is in contrast to the emergent physicalist who only extends such states to the parts of their surroundings which they consider to be sufficiently similar to themselves. However, the panwhat-it-is-likeness physicalist *does not* extend states of awareness to all parts of the physical. In contrast to this view the panawareness physicalist extends not only what-it-is-likeness to all of the physical but also extends awareness to all of the physical. This is typically because the panawareness physicalist believes that wherever there is what-it-is-likeness there is awareness.[23] The issue which separates the panwhat-it-is-likeness physicalist from the panawareness physicalist – whether what-it-is-likeness and awareness come apart – is explored in *Chapter Four* and *Chapter Six*.

There is a long tradition according to which the entire universe is pervaded with awareness (I am referring to this view as 'panawareness physicalism'). Philosophers in this tradition are often referred to as panpsychists or panexperientialists.

---

[23] Indeed, this is a very widely held belief, as we will see in *Section 6.1*. Because of this panawareness can be thought of as a variety of panwhat-it-is-likeness. That is to say, the panawareness advocate agrees with the panwhat-it-is-likeness advocate that the physical is pervaded with what-it-is-likeness, but makes an additional claim – wherever there is what-it-is-likeness there is awareness. In future sections and chapters the term panwhat-it-is-likeness will sometimes be used to refer to the general claim that all of the physical is pervaded with what-it-is-likeness (panwhat-it-is-likeness and panawareness) and sometimes it will be used in the narrow sense (as being opposed to panawareness). It should be clear in which sense the term is being used.

However, this terminology is potentially confusing. The panpsychist believes that 'psyche' pervades the universe, but there are many possible conceptions of what exactly 'psyche' is. Indeed, one could believe that *any* type of interiority is sufficient for 'psyche'; if so, then the panwhat-it-is-likeness physicalist will also be a panpsychist.[24] The panexperientialist believes that 'experience' pervades the universe (but *not* 'psyche'). But what is experience? How is experience different from psyche? How is panpsychism different from panexperientialism?

There actually appears to be no intelligible difference between these two positions. The term panexperientialism was simply created to sound *less audacious* than panpsychism, as we will see in a moment. Furthermore, panexperientialism entails that all of the physical is aware – so panexperientialism is a variety of panawareness physicalism. This implies that panpsychists also typically hold that all of the physical is aware; for, to hold that all of the physical is aware is *more audacious* than to hold that all of the physical is not aware, and panexperientialists hold that all of the physical is aware.[25] This is clear in the following two assertions from panexperientialist David Ray Griffin (2007, p.78, pp.130-1), the first of which attempts to justify his invention of the term 'panexperientialism':

---

[24] So, the term 'panpsychism' seems to be just as unhelpful as the term 'physicalism'. In contrast, it is (or at least will become) obvious what my terms 'panawareness' and 'panwhat-it-is-likeness' mean.

[25] However, as I have already noted, the term 'panpsychism' is a vague term and actually covers a range of positions depending on how one conceptualises 'psyche'.

"Panpsychism" is the term that has generally been used for this position. "Panexperientialism" is preferable, however, for two reasons: (1) The term "psyche" suggests that the basic units endure through long stretches of time, whereas they may be momentary experiences; and (2) "psyche" inevitably suggests a higher form of experience than would be appropriate for the most elementary units of nature.

experience always involves some minimal awareness of *what is*

It seems that Griffin's use of the term 'panexperientialism' rather than 'panpsychism' is partly motivated by the over-the-top reactions that some humans have to the word panpsychism. For example, McGinn (1982, p.32) claims that: "[Panpsychism] is metaphysically and scientifically outrageous." Perhaps Griffin's hope is that McGinn would not consider panexperientialism to be quite so outrageous. However, the point is that panexperientialism is not a different position from panpsychism as panpsychists also agree with Griffin's two assertions above, that: (1) the basic units may be momentary experiences and (2) the basic units have a lower form of experience than certain more complex arrangements of such units. This is clear in the following quotes from panpsychist advocates David Bohm, Galen Strawson and Ervin Laszlo:

in some sense, a rudimentary mind-like quality is present even at the level of particle physics.

(Bohm, 1990)

we will have to wonder how macroexperientiality arises from microexperientiality, where by microexperientiality I mean the experientiality of particles relative to which all evolved experientiality is macroexperientiality.[26]

(Strawson, 2006, p.26)

psyche is [not] present throughout reality in the same way, at the same level of development. We [panpsychists] say that psyche evolves, the same as matter.

(Laszlo, 2004, p.147)

So, it is clear that the idea that there are different levels of experience=awareness within the universe is a crucial element in both panpsychism and panexperientialism. It is this idea which gives rise to the panawareness 'combination problem'. William Seager (1995) describes this problem as follows: "explaining how the myriad elements of 'atomic consciousness' can be combined into a new, complex and rich consciousness such as that we possess."[27] Put slightly differently, the 'combina-

---

[26] As we will see in *Chapter Six*, Strawson, like most people, takes the term experience to entail awareness.

[27] The concept of consciousness will be explored in the next three chapters. I take it to be obvious that in quotes such as these the word 'consciousness' is

tion problem' is the problem of how to give an intelligible account of how 'micro' experiences=awarenesses sum to form a 'macro' experience=awareness. Formulating such an account is no easy task. William James (1950, p.160) argued that such 'combination' is logically untenable:

> Take a hundred of them [feelings], shuffle them and pack them as close together as you can (whatever that might mean); still each remains the same feeling it always was, shut in its own skin, windowless, ignorant of what the other feelings are and mean. There would be a hundred-and-first feeling there, if, when a group or series of such feelings were set up, a consciousness *belonging to the group as such* should emerge. And this 101st feeling would be a totally new fact; the 100 feelings might, by a curious physical law, be a signal for its *creation*, when they came together; but they would have no substantial identity with it, nor it with them, and one could never deduce the one from the others, or (in any intelligible sense) say that they *evolved* it.

Despite believing that such combination from micro-experience=awareness to macro-experience=awareness is logically untenable James ultimately concluded that it must somehow occur. Similarly, in considering the varieties of physicalism, Strawson (2009, p.64) claims that "some sort of

---

being used to refer to awareness. So, to talk of 'atomic consciousness', and to believe that 'atomic consciousness' pervades the physical, is to be a panaware-ness physicalist.

panpsychism must be true" but admits that he has no account of combination; indeed, Strawson (2009, p.64) "enthusiastically agree[s]" with Philip Goff's (2006a, p.26) assertion that Strawson simply has "faith that it must happen somehow". In short, the panawareness physicalists James and Strawson believe that combination must occur but they cannot explain how it could possibly occur.

3.5.5 *Dualism*

In the previous few sections I have outlined several contemporary world views which are grounded in the belief that the universe is comprised of one type of stuff; the word 'physical' has been used to refer to this 'one type of stuff'. We have explored three types of physicalism – 'emergent physicalism' which entails that this 'one type of stuff' is largely devoid of interiority; 'panwhat-it-is-likeness physicalism' which entails that this 'one type of stuff' is pervaded with 'what-it-is-likeness'; and 'panawareness physicalism' which entails that this 'one type of stuff' is pervaded with awareness='what-it-is-likeness'.

The existence of these varieties of physicalism makes the term rather redundant – it is just a placeholder for the belief that the universe is comprised of one 'type of stuff' the nature of which is unknown. The alternative view is that the universe is comprised of two 'types of stuff'; this view is 'dualism'.[28] There is a curious thing about 'dualism'. At first glance it could be thought of as the complete opposite of physicalism; however, on

---

[28] Of course, one could also believe that the universe is comprised of 3, or 4, or 5, or 189 types of stuff. The analysis of 'dualism' will also cover these possibilities. I am using the term 'dualism' to encompass the positions which are referred to as 'property dualism' and 'substance dualism'.

closer inspection there seems to be no intelligible difference between dualism and physicalism. For, as we have seen, the nature of the physical is fundamentally mysterious to us – there are unknown physical properties. The term 'physicalism' is actually just a placeholder to refer to the belief that there is 'one type of stuff' of which we know not. What this means is that when a 'dualist' asserts that there are two 'types of stuff' in the universe this is perfectly compatible with 'physicalism'; for, the properties that the 'dualist' takes to be a second 'type of stuff' could be an unknown aspect of the physicalists' singular 'type of stuff'. So, there is a linguistic and conceptual difference between uttering 'one' and 'two', but when one accepts that one doesn't know what the 'one' refers to then one has to conclude that there is no meaningful distinction between 'physicalism' and 'dualism'.

Why does the 'dualist' feel the need to assert that there is a second 'type of stuff' in humans? The 'dualist' seems to be motivated to make this assertion by their belief that humans are very different from their surroundings. This belief is *also* at the core of 'emergent physicalism'. It is only within 'panawareness physicalism' and 'panwhat-it-is-likeness physicalism' that the hypothesised chasm between humans and their surroundings is rejected.

Furthermore, when it has been acknowledged that full knowledge of the nature of the physical is barred from humans – that humans do not know what the physical is – then it is far from clear what it means to talk of the non-physical. Therefore, all of the 'physicalist' positions outlined in the previous three sections can be relabelled as 'dualist' positions. When this is done one can then say that these positions contain two 'types of stuff' rather than one; however, the claims that the various positions make concerning the nature of the universe is unchanged!

So, rather than being an 'emergent physicalist' one could be an 'emergent dualist' and claim that the 'interiority' properties which emerge in some parts of the universe are 'dualistic' rather than 'physicalistic' properties. In this vein the 'dualist' Richard Swinburne (1988, p.1) claims that:

> the fact of evolution is evident. Even more evident, to my mind, is the fact that what has evolved is different, radically and qualitatively, from that from which it has evolved. Rocks and rivers are not conscious; they do not have thoughts, sensations and purposes; but men, and some animals, do have thoughts, sensations, and purposes.

The first thing to note about this assertion is that Swinburne's speculations are just that – speculations. He doesn't know whether what has evolved is actually "radically and qualitatively" different (yet he calls this speculation a 'fact'!). Despite believing that evolution is "evident" his thinking still seems to be firmly grounded within a 'static' view of the universe.

Swinburne believes that the evolved 'dualist properties' are of a "radically" different nature to 'physical properties'. However, when one realises that one doesn't know the nature of 'physical properties', then what this belief amounts to is unclear. It seems to amount something like the following: *X is radically different to Y, but I know very little about Y.* This kind of thing shouldn't convince anyone. Furthermore, whichever property the 'emergent dualist' takes to be a second 'type of stuff', the 'emergent physicalist' can claim to be an evolved 'monist physical property'. For example, if the 'emergent dualist' claims that there is a part of humans which continues to exist following the death of their biological body, then the 'emergent physicalist' can agree and claim that this part is part of the 'monist physical stuff'.

Another possibility for the 'dualist' is to agree with the 'panwhat-it-is-likeness physicalist' that all of the physical is pervaded with 'what-it-is-likeness', and that awareness is something which only emerges in parts of the universe. The 'panwhat-it-is-likeness dualist' can then disagree with the 'panwhat-it-is-likeness physicalist' about the nature of awareness; whilst the 'panwhat-it-is-likeness physicalist' claims that awareness is 'physical', the 'panwhat-it-is-likeness dualist' claims that it is 'non-physical'.

The third possibility for the dualist is a modified version of 'panawareness physicalism'. On this account the 'panawareness dualist' asserts that a human has a 'physical' part which is wholly devoid of both awareness and 'what-it-is-likeness', and a 'non-physical' part which contains awareness='what-it-is-likeness'. However, this applies not just to humans, for according to the 'panawareness dualist' *the entire* universe has both of these 'types of stuff'.

3.6 Conclusions

In this chapter we have covered a lot of ground. We have considered how the relationship between a human and their surroundings leads to particular conceptions of those surroundings being formed. We have seen that a human's perceptions can be conceptualised in many different ways resulting in different world views. Furthermore, we have seen that a human's perceptions are inevitably constrained in several different ways. This means that human conceptualisations of their surroundings are grounded in inevitably constrained perceptions.

The result of all of this is that there are many diverse world views concerning how similar humans are to their surroundings. We have considered several contemporary world views – physicalism, emergent physicalism, panwhat-it-is-likeness physicalism, panawareness physicalism and dualism. I have tried to convince you that the term 'physical' is merely a placeholder for the belief that there is only one 'type of stuff' in the universe, and that there is no meaningful distinction between physicalism and dualism. I have suggested that those who do not acknowledge the existence of the inevitable constraint are more likely to believe that humans are dissimilar to the vast majority of their surroundings. On the contrary, those who are embedded within the evolutionary perspective, and thereby recognise the existence of the inevitable constraint, are likely to believe that their surroundings are quite similar to themselves.[29] The 'dualist'/'emergent physicalist' believes that humans are very dissimilar to the surroundings which evolved them; whilst the 'panwhat-it-is-likeness physicalist' and the 'panawareness physicalist' believe that humans are very similar to the surroundings which evolved them.

From the evolutionary perspective there are very good reasons to believe that *in reality* humans are very similar to the

---

[29] One can realise that the inevitable perceptual constraint exists without being embedded in the evolutionary perspective. My claim is simply that if a human is embedded within the evolutionary perspective that they are much more likely to realise that the inevitable constraint exists.

surroundings which evolved them. If everything that now exists has evolved from the Big Bang; if the universe is a slowly evolving interconnected whole; then there is a good reason to believe that there is not a great chasm between humans and their surroundings. As we saw in *Chapter One* there is a good reason to believe that most, if not all, of the attributes of humans are likely to be pervasive in the universe.

I have tried to understand why contemporary humans ordinarily assume that their surroundings are very dissimilar to themselves, and have suggested that this is because the majority of humans are embedded within the 'static view' and so consider the relationship between themselves and their surroundings purely in the present. A corollary of this, which provides a further explanation, is that the majority of humans do not realise the range of ways in which their perceptual apparatus is inevitably constrained.

When you have digested all of this I believe that there is a good chance that you will reject the idea that there is a great chasm in the universe, and that you will therefore reject 'emergent physicalism' which claims that the vast majority of the universe is devoid of interiority. This means that you are likely to be attracted to a different 'world view' such as 'panwhat-it-is-likeness physicalism' or 'panawareness physical-ism'. In order to get a greater appreciation of what these two positions entail we need to consider various other aspects of the relationship between humans and their surroundings. In the next chapter we will consider the phenomena of consciousness and mind. In *Chapter Five* we will consider the phenomena of 'what-it-is-likeness' and the human senses. And, in *Chapter Six,* we will consider the relationship between awareness, perception and 'what-it-is-likeness'.

Chapter 4

Consciousness and mind

In the previous chapter we considered how humans conceptualise the relationship between themselves and their surroundings. We considered two broad world views – the Primal World View and the Modern World View, and several other world views – physicalism, panwhat-it-is-likeness, panawareness and dualism.[30] There is a common thread which runs through all of these various 'world views': consciousness and mind. What I mean by this is that the particular conceptualisation that a human has of mind and/or consciousness is likely to play a very large role in determining which 'world view' they adhere to. In other words, the concept of 'having a mind', and/or of 'being conscious' is one of the key ways that humans typically seek to carve a division in the universe between themselves and their surroundings, or between themselves and a few other species of animal, and the rest of the universe.

My aim in this chapter is to explore the concepts of consciousness and mind. This is important because there are clearly 'thick' and 'thin' concepts of both mind and consciousness. By this I mean that one can have a conceptualisation of

---

[30] From now on I will refer to 'emergent physicalism' as 'physicalism', 'panwhat-it-is-likeness physicalism' as 'panwhat-it-is-likeness', and 'panawareness physicalism' as 'panawareness'. I generally won't refer to 'dualism' due to the conclusion, in *Chapter Three,* that this position is not different from the three other positions I have just mentioned.

mind or consciousness in which a lot of different attributes are 'packed in' to the concept, or one can have a very simple conceptualisation which contains a single attribute. Clearly, if one has a 'thick' conceptualisation of mind (or consciousness) *and* believes that only humans have a mind (or consciousness) then one is likely to believe that there is a great chasm between humans and their surroundings. On the contrary, if one has a 'thin' conceptualisation of mind (or consciousness) *and* believes that only humans have a mind (or consciousness) then one will believe that there is *a difference* between humans and their surroundings, but one won't necessarily believe that there is a chasm. So, attempting to get an understanding of what it means to be conscious, and what it means to have a mind, is a crucial part in our continuing exploration into the nature of the relationship between humans and their surroundings. In *Section 4.1* I explore what it means to be conscious by considering the question: What is the problem of consciousness? In *Section 4.2* I explore what it means to have a mind. Finally, in *Section 4.3*, I draw some conclusions.

4.1   What is the 'problem of consciousness'?

The 'problem of consciousness' is widely taken to be the problem of accommodating consciousness within a physicalistic/materialistic worldview. This problem is widely seen as the biggest challenge humans face as they seek to gain a more comprehensive understanding of both themselves and their surroundings. But what exactly is consciousness?

I contend that the 'problem of consciousness' is widely conceived of as the problem of how it is possible that states of 'what-it-is-likeness' can exist in a physical world. I propose that it is this conceptualisation of the problem which leads to its seeming intractability and outline an alternative conceptualisation in which there are two distinct problems – the *problem of awareness* and the *problem of 'what-it-is-likeness'*. I suggest that this alternative conceptualisation largely demystifies the 'problem of consciousness'. Furthermore, this alternative conceptualisation entails a 'thin' concept of consciousness which can forge a world view in which there is no great chasm between humans and their surroundings. Here is the standard contemporary view of the problem of consciousness:

> Both in philosophy and psychology "the problem of consciousness" is supposed to be very special… It has to do with the internal or subjective character of experience, paradigmatically sensory experience, and how such a thing can be accommodated in, or even tolerated by, a materialist theory of the mind.
>
> William G. Lycan (1996, p.1)

> Consciousness is widely regarded as an intractable mystery… By what mysterious power do our material brains generate these additional conscious feelings?
>
> David Papineau (2004, pp.1-2)

> Human consciousness is just about the last surviving mystery.
>
> Daniel Dennett (1991, p.21)

we know that brains are the *de facto* causal basis of consciousness, but we have, it seems, no understanding whatever of how this can be so. It strikes us as miraculous, eerie, even faintly comic... we are cut off by our very cognitive constitution from... [conceptualizing] the psychophysical link.

Colin McGinn (1989)

The common theme underlying these views is that accommodating consciousness within a materialistic worldview is the biggest challenge humans face as they seek to gain a 'complete' understanding of both themselves and their surroundings. According to Lycan this "problem of consciousness" is "very special"; according to Papineau the problem is widely regarded as an "intractable mystery"; according to Dennett the problem is "just about the last surviving mystery". McGinn goes further and claims that humans are incapable of solving the problem; however, despite this he curiously claims that humans *can* know that brains cause consciousness. Why is the "problem of consciousness" so special and so mysterious? Answering this question requires a clear understanding of what is meant by the term 'consciousness'.

There is a standard contemporary philosophical use of the term consciousness, a use that is revealed in the above quotes. Papineau describes the problem of consciousness as the problem of how material brains generate conscious "feelings", whilst Lycan describes the problem as to do with the "subjective character of experience, paradigmatically sensory experience". The standard conceptualisation of the phenomena of consciousness is that the term 'consciousness' refers to states that *feel* a certain way, states that have a certain *subjective character*.

These states are also widely referred to either as states of 'what-it-is-likeness' or as states of qualitative feeling/qualia. This standard conceptualisation of consciousness is also clear in the following assertions made by David Chalmers, John Searle and Galen Strawson:

> We can say that a being is conscious if there is *something it is like* to be that being.
>
> (Chalmers, 1996, p.4)

> It is a remarkable fact about consciousness... that there is a qualitative feel to any conscious state.
>
> (Searle, 2005, p.202)

> 'consciousness', conscious experience, 'phenomenology', experiential 'what-it's likeness', feeling, sensation, explicit conscious thought as we have it and know it at almost every waking moment. Many words are used to denote this necessarily occurrent (essentially non-dispositional) phenomenon, and I will use the terms 'experience', 'experiential phenomena' and 'experiential-ity' to refer to it.
>
> (Strawson, 2006, p.3)

So, despite the use of slightly different terminology – "something it is like", "feeling", "qualitative feel", "experiential 'what-it's likeness'", "subjective character of experience", "phenomenology" – it is clear that the "problem of consciousness" is widely taken to be the problem of how it is possible that states that feel a certain way can exist in a material world; for simplicity I will use just one term 'what-it-is-likeness' to refer to these states. So, the "problem of consciousness" on the standard contemporary view is the *problem of 'what-it-is-likeness'*. That this is so should already be clear, but let us also consider how McGinn (1999, p.15, pp.2-3) describes the "problem of consciousness":

> We have a good idea how the Big Bang led to the creation of stars and galaxies, principally by the force of gravity. But we know of no comparable force that might explain how ever-expanding lumps of matter might have developed an inner conscious life.

> consciousness... the having of sensations, emotions, feelings, thoughts.

So, according to McGinn, the "problem of consciousness" is that we have no idea what the "force" is that could cause matter to "develop an inner conscious life". And according to McGinn the term 'consciousness' means sensations, emotions, feelings and thoughts. So, whilst McGinn uses the term to refer to a diverse array of phenomena (sensations, emotions, feelings and thoughts) most, if not all, of these phenomena are states that feel a certain way, they are states of 'what-it-is-likeness'. This means that McGinn also takes the "problem of consciousness" to be the problem of how it is possible for states of 'what-it-is-likeness' to exist in a material world.

What have we concluded so far? We have concluded that the "problem of consciousness" is widely conceived of as the problem of how it is possible for states of 'what-it-is-likeness' to exist in a physical world. And, there are widely conceived to be a large number of 'what-it-is-likeness' phenomena, such as thoughts, feelings, emotions and sensations.

Let us imagine that the universe is wholly comprised of states of *'nothing-it-is-likeness'* (which, of course, it isn't) then, according to the standard view there would seemingly be no "problem of consciousness" in need of a solution. In other words, on the standard view it is the *mere existence* of states of 'what-it-is-likeness' which generates the "problem of consciousness"; why particular material states are (or are associated with) particular 'what-it-is-likeness' states is a separate issue. In short, if one is a materialist *and* one asserts that there was a time when the material world was wholly devoid of 'what-it-is-likeness' then one faces the "very special" problem of having to explain how that world evolves 'what-it-is-likeness'. According to the standard contemporary view *this problem* is the "problem of consciousness".

A seeming implication of the standard contemporary view is that if one were to suppose that the entire material world is, and always has been, pervaded by states of 'what-it-is-likeness' (this is the panwhat-it-is-likeness view which we encountered in *Chapter Three*) then the "problem of consciousness" would no longer exist. However, we have already seen that the panwhat-it-is-likeness advocate *does* face the problem of seeking to explain how awareness arises out of that which is wholly unaware.

What does this mean? It means that there is an implicit assumption underpinning the standard contemporary view of consciousness; it is assumed that to talk of consciousness is to talk of *both* 'what-it-is-likeness' and 'awareness'. So, when the advocates of the standard contemporary view refer to 'what-it-is-likeness' they are assuming that states of 'what-it-is-likeness' *are* states of awareness. They assume that if there is a state of awareness that this state *itself* feels a certain way, and they assume that if there is a state that feels a certain way that this state *itself* must be a state of awareness. On this view, if one has an account of 'what-it-is-likeness' then one also has an account of awareness.

Now, if panwhat-it-is-likeness is an intelligible position, which it surely is, then this means that the "problem of consciousness" is actually two distinct problems – the *problem of 'what-it-is-likeness'* and the *problem of awareness*. In conflating the two problems the standard contemporary view takes itself to be addressing the *problem of awareness* by addressing the *problem of 'what-it-is-likeness'*. However, if there are actually two problems then this is clearly inappropriate. For, the 'panwhat-it-is-likeness' advocate does not face the *problem of 'what-it-is-likeness'* but they still face the *problem of awareness*. When the *problem of 'what-it-is-likeness'* is solved the "problem of consciousness" can remain unsolved.

What exactly is the *problem of awareness*? As a first insight into the nature of this problem we can consider the following passage from Iris Murdoch's (1973, p.224) *The Black Prince*:

> Angels must wonder at these beings who fall so regularly out of awareness into a fantasm-infested dark. How our frail identities survive these chasms no philosopher has ever been able to explain.

Now, the beings in question are clearly human beings, and our current interest is not the existence of human personal identity through time, it is the related issue of such identities being "frail" because we are beings who "fall so regularly out of awareness" into "chasms" of darkness. In short, humans are parts of the universe which oscillate between having awareness and losing awareness. Of course, there will be some humans who deny this, but to the overwhelming majority of humans it is surely an obvious fact that they, as humans, oscillate between having awareness and losing awareness. Not only does this seem to be an obvious fact, it is also seems to be one of the most fundamental aspects of human existence. At a particular time after being forged from a zygote a human first becomes aware. Then at a future time this awareness is lost as the human enters a state that is wholly devoid of awareness (most humans refer to this state as an 'unconscious' state or as a 'sleep' state). From this moment on the rest of a human life is shaped around the *seemingly obvious fact* that a human repeatedly oscillates between states of awareness and states where this awareness is lacking. The life of a typical human ends when this oscillation cycle ceases, when a state of unawareness is entered from which there is no return.[31]

If one accepts that awareness of the universe is something which has evolved – that if one goes far enough back in time that the entire universe was wholly devoid of awareness – then clearly one faces the *problem of awareness*. This is the problem of explaining how it is possible that a universe which is devoid of awareness can evolve awareness.

---

[31] This oscillation cycle is explored further in *Section 6.6*.

Of course, according to the panawareness view, which we encountered in *Chapter Three,* the entire universe is, and always has been, pervaded by states of awareness. We saw that the panwhat-it-is-likeness advocate believes that when two stones collide this involves states of 'what-it-is-likeness' but that it does not involve states of awareness.[32] On the contrary, the advocate of panawareness believes that this interaction involves both 'what-it-is-likeness' and awareness ('what-it-is-likeness'=awareness). This means that the panawareness view is in opposition to both the standard contemporary view and panwhat-it-is-likeness. For, according to both of these views awareness is a state of the universe which evolves – it is just that the standard contemporary view asserts that this evolutionary event is simultaneously the evolution of 'what-it-is-likeness', while the panwhat-it-is-likeness advocate denies that this is so.

What *is* the "problem of consciousness"? According to the standard contemporary view it is the problem of explaining how states of *'what-it-is-likeness'=awareness* evolve in a physical universe. According to the panwhat-it-is-likeness advocate it is the problem of explaining how states of *awareness* evolve in a panwhat-it-is-likeness universe. The panawareness advocate could deny that there is a problem in need of solution; for, they could hold that the entire universe is continuously aware and always has been; however, this clearly does not sit well with the *seemingly obvious fact* that awareness is a state of the universe that pops into and out of existence. The other option open to the panawareness advocate is to hold that awareness exists at different levels; they could hold that a human has awareness,

---

[32] It is worth stressing this point because many humans have an ingrained way of thinking according to which to talk of 'what-it-is-likeness' is to talk of awareness.

and that the parts of the universe that constitute a human also have awareness; in this way the panawareness advocate can accept the *seemingly obvious fact*.[33] If the panawareness advocate takes this approach then they will face the 'combination problem' – the problem of explaining how states of '*what-it-is-likeness*'=*awareness* at a low level combine to form states of '*what-it-is-likeness*'=*awareness* at a high level.[34]

So, the "problem of consciousness" is the *problem of awareness*, the *problem of* '*what-it-is-likeness*'=*awareness*, or the '*combination problem*'. There is a strong case for adopting the panwhat-it-is-likeness view. This view fully naturalises 'what-it-is-likeness' and takes much of the mystery out of the "problem of consciousness". It is only if one believes that the material world is largely devoid of 'what-it-is-likeness' that the "problem of consciousness" (*the problem of* '*what-it-is-likeness*'=*awareness*) is "miraculous" (McGinn) and "very special" (Lycan). If one takes the panawareness route then one either faces the equally problematic 'combination problem' or one has to deny the *seemingly obvious fact*. On the panwhat-it-is-likeness view the *problem of consciousness* still needs to be addressed. However, this problem is not particularly miraculous or special; it is simply the *problem of awareness*. According to this view to be conscious is simply to be aware.[35]

---

[33] The panawareness advocate could believe that there are different levels of awareness within a human and also deny the seemingly obvious fact.

[34] The 'combination problem' was considered in *Section 3.5.4.*In *Section 6.8* I reject the idea that there are 'levels' of awareness.

[35] I am contending that 'what-it-is-likeness' pervades the universe but that awareness does not. You might be wondering if awareness *itself* is wholly devoid of 'what-it-is-likeness'; in *Section 6.3* I will contend that it is.

We will consider the nature of 'what-it-is-likeness' in *Chapter Five;* in *Chapter Six* we will both consider the nature of awareness and the relationship between awareness and 'what-it-is-likeness'. For now, let us consider various conceptions of the mind.

4.2 What is a Mind?

The concept of 'having a mind' is one of the key concepts that humans use to carve a division of extreme significance into the universe. Humans say that parts of the universe 'have minds' and that parts of the universe don't 'have minds'. But, what exactly is a mind?

The aim of this section is not to consider whether a computer has a mind, whether a tree has a mind, or whether an atom has a mind; our aim is simply to reflect upon the question of what it means for a mind to exist. To put this slightly differently, our objective is *not* to consider whether there actually is a distinction in the universe between the minded and the non-minded, it is simply to consider what it could possibly mean to assert that such a distinction exists. This is important because, as with consciousness, there are both 'thick' and 'thin' conceptions of what 'having a mind' entails, and the conception that one has is likely to influence how one envisions the relationship between humans and their surroundings.

It is far from clear what it means to 'have a mind'; that is to say, it is far from clear what it means to say that part of the universe is 'mental' whilst part of the universe is 'non-mental'.[36]

---

[36] To be clear, if a 'mental' attribute exists then 'a mind' exists. If a particular attribute exists but 'a mind' doesn't exist then this attribute is clearly not a 'mental' attribute.

According to Herbert Feigl the term 'mental' is precariously vague which means that progress in the mind—body problem requires a study of all of the attributes that are often taken to be 'mental'. Feigl (1967, p.20, pp.45-6) claims that:

> The terms "mental" and "physical" are precariously ambiguous and vague. Hence a first prerequisite for the clarification and the adequate settlement of the main issues [in the mind-body problem] is an analytical study of the meanings of each of the two key terms

> the term "mental" in ordinary and even scientific usage represents a whole family of concepts...special distinctions like "mental$_1$", "mental$_2$", "mental$_3$", etc. are needed in order to prevent confusions

In accordance with this view the purpose of this section is to approach the question of what a mind is through considering a number of attributes which are often thought to be mental attributes. Firstly, the idea that a mind is a central core of human attributes is considered and rejected. Then, most of the section is spent exploring the idea that particular attributes are mental attributes; many of these supposedly 'mental' attributes are considered.

Our approach is as follows. Let us suppose that a particular 'mental' attribute exists; is one inclined to assert that its existence means that a mind exists? If not, then the attribute is seemingly not sufficient for the existence of a mind; there is, or could conceivably be, a part of the universe in which this attribute exists but in which a mind does not exist. The converse supposition will also be used: let us suppose that this particular attribute does not exist; is one inclined to assert that its absence means that a mind does not exist? If not, then the attribute is seemingly not necessary for the existence of a mind. This methodology – a conceptual analysis of the judgements about what seems to be obvious – is the only way that we can make progress concerning what it means to 'have a mind'.

So, our objective is to identify which attributes humans intuitively hold to be necessary and/or sufficient for the existence of a mind. In this way our aim is to cut through the "precarious ambiguity and vagueness" of mental talk and get to the heart of what it really means for a mind to exist.

4.2.1 A central core of human attributes?

When one reflects on what a mind is one of the first conclusions one will reach is surely that it is an attribute that at least most humans possess. This starting 'conclusion' is unproblematic; whatever having a mind entails it is indubitably the case that at least most humans possess this attribute. This is so because 'mind' is a term invented by humans to refer to an attribute or attributes that at least most humans possess! However, some humans *start out* with a list of attributes that they take to be central to *being a human*, and then argue that having these attributes is 'having a mind'. This *is* deeply problematic. For example, Eric Matthews, D. M. Armstrong, and Richard L. Gregory state that:

To say what we mean by the mental is... a matter of distinguishing a central core of human attributes, activities, processes and so on

(Matthews, 2005, p.51)

Men have minds, that is to say, they perceive, they have sensations, emotions, beliefs, thoughts, purposes and desires. What is it to have a mind? What is it to perceive, to feel emotion, to hold a belief or to have a purpose?

(Armstrong, 1980, p.1)

mind a vague term, covering control of intentional behaviour and awareness...learning, memory, perception, emotion, intention, aesthetics, and much more

(Gregory, 2004, p.xviii)

This conception of what having a mind entails is problematic because if one generates one's conception of 'having a mind' by formulating a list of attributes that humans have, then the question is: How exactly does one derive the right list? The overwhelming majority of humans do not believe that trees, stones and tables 'have a mind'. As we saw in *Chapter One* and *Chapter Three,* there is a widespread human tendency to believe that the majority of human surroundings are most un-humanlike. What this means is that when a typical human is formulating their list of the 'central core' of human mental attributes, they are likely to derive their list on the basis of which attributes they believe that humans possess but that stones and tables lack. Once this 'central core' of human attributes has been created it is then asserted that these attributes – attributes which stones and tables are believed to lack, but which humans don't lack – are 'mental' attributes. Given the limitations of the human perceptual apparatus this is a very problematic way of attempting to carve a division in the universe between the 'mental' and the 'non-mental'.

Those who reject the 'central core' view will rightly ask for justification as to why exactly this list of attributes constitutes a mind rather than a slightly different list. Why not a longer list? Why not a shorter list? In other words, the objection is that the list is wholly arbitrary. Fred might have a list of five attributes, Sally might have a list of four attributes, and Tom might have a list of ten attributes. Furthermore, every few days Tom might change his mind; one day his 'central core' of mental attributes has ten members, a few days later it has increased to eleven, and a week later it has decreased to seven! This is clearly a hopeless approach to the question of what it means to 'have a mind'.

Fred asserts that the feeling of pain is a mental attribute; Tom who is a panwhat-it-is-likeness advocate denies that this is so. Sally believes that intentionality is a mental attribute; Fred, who has been contemplating Molnar's views on 'physical intentionality', denies that this is so.[37] Tom insists that perception is a mental attribute; Sally, who is a panperceptualist, denies that this is so. If we are to make progress, and to cut through the 'precariously ambiguous and vague' nature of mental talk, we clearly need to consider individual attributes

---

[37] Molnar's view, that intentionality is more pervasive than the mental, is discussed in *Section 4.2.5.*

and whether or not they are necessary and/or sufficient for the 'having of a mind'.

4.2.2 Freedom?

> The second volume of *The Life of the Mind* will be devoted to the faculty of the Will and, by implication, to the problem of Freedom
>
> (Arendt, 1978, p.3)

Could there be a single attribute which if it exists means that 'a mind' exists? Our concern in this section is the belief that if freedom exists then this means that 'a mind' exists; indeed, freedom is often taken to be the hallmark of 'a mind'. What exactly is freedom? Sometimes the concept of freedom is envisioned as freedom from constraint; on this view a human who is put in a straightjacket and kept in solitary confinement is devoid of freedom. This is not the conception of freedom which one is referring to when one asserts that freedom is the hallmark of a mind. One would not say that a particular human had a mind at t_1, and a moment later at t_2 was devoid of a mind solely because they had been put into a straightjacket. Rather, to say that freedom is the hallmark of a mind is to envision that a part of the universe has *alternative possible future states even if that part encountered identical future surroundings*. In other words, a part of the universe lacks freedom if its states are *wholly determined* by its surroundings, and a part of the universe has freedom if its states are self-determining *despite* its surroundings.

Now, it is a trivial truth that one has no way of knowing whether all of the states in the universe are wholly determined by their surroundings. The entire universe could be pervaded by freedom; contrarily, the entire universe could be wholly devoid of freedom.[38] So, it is possible that the entire universe could be wholly devoid of freedom, a scenario which entails that one, one's family and lifelong friends, and all other humans, are wholly devoid of freedom. If one accepted that this was so, would one then feel inclined to assert that all humans lacked a mind? This conclusion would be a most bizarre one to reach; one would surely conclude that humans have a mind irrespective of whether or not they have freedom. Therefore, the only reasonable conclusion to reach is that freedom is not necessary for the existence of a mind because minds exist even if universal determinism is true; so, freedom is not a 'mental' attribute.

4.2.3 Feeling States?

> Many people who philosophise about the mind, interestingly, in giving an example of something 'mental', tend to concentrate on pain sensations... But they do not seem to be central to what we think of as 'mental'.
>
> (Matthews, 2005, p.50)

---

[38] If one has trouble accepting the possibility that the entire universe could be pervaded with freedom, then it might be helpful to recall the discussion of the inevitable constraints in *Section 3.1*. In particular, the constraint of importance is the temporal constraint which can cause humans to conclude that things are determined when they could actually be long-term expressions of freedom.

Animals, whether or not they can reason, certainly have minds in the sense of having feelings of pain and pleasure

(Matthews, 2005, p.76)

states of mind – headaches

(Heil, 1998, p.61)

As Eric Matthews notes, many people who philosophise about 'having a mind' believe that feeling states such as pain and pleasure are paradigmatic mental attributes. Matthews claims that such states are *not central* to 'having a mind'. The question before us here is: Does the existence of such states have *anything* to do with having a mind? The proposed answer is that the existence of a feeling state has nothing whatsoever to do with having a mind.

Let us consider the possibility – which is a reality if panwhat-it-is-likeness is true – that when two hydrogen atoms and an oxygen atom interact to form a water molecule a certain feeling state is generated; let us refer to this particular feeling state as an 'htwoain' state. If this strikes one as implausible one should keep in mind that panwhat-it-is-likeness does not entail *either* that the existence of a feeling state is accompanied by awareness of that state, *or* that a feeling state is accompanied by a judgement as to whether the state is 'good' (pleasurable) or 'bad' (painful).

Now, the issue before us is as follows: if a solitary 'htwoain' feeling state existed would one feel inclined to assert that a mind exists? As already stated, there is no awareness of the existence of the state or any thought concerning the state – that it is either painful or pleasurable; there is just a tiny part of the universe which is a particular feeling state. The answer is, surely, that one would not feel inclined to assert that a mind exists; whatever a mind is it surely must be more than this.[39]

A less abstract example than the 'htwoain' scenario is the possibility of the existence of bodily pains which exist outside of the awareness of the human whose body those pains are located in. According to David Rosenthal (1997, p.732):

> If we are intermittently unaware of a pain by being distracted from it, we feel the pain only intermittently; similarly with its hurting and our being in pain. Still, it is natural to speak of having had a pain that lasted throughout the day, and even to say that one was not always aware of that pain. This provides evidence that commonsense countenances the existence of nonconscious pains. Feeling pains and having them seem equivalent only because of our lack of interest in the nonconscious cases.

In this 'pain without awareness' scenario feeling states exist in the absence of the awareness of their existence. A similar question arises as with the 'htwoain' scenario: Would one want to assert that the existence of a solitary feeling state in a human

---

[39] There are always exceptions to the rule. I take this conclusion to be the one which the overwhelming majority of humans would reach. There are a small minority of humans – some panpsychists – who believe that all atoms are states analogous to 'htwoain' states; they take this to entail that all atoms are minds.

body which exists without any awareness of its existence, but which one would judge as 'painful' if one was aware of it, entails the existence of a mind? Again, I take it that the answer is obviously no. To say that such a state exists and to say that a mind exists is to say two entirely different things.

Let us consider whether 'a mind' can exist in the total absence of feeling states. Let us imagine that a human is wholly devoid of feeling states, but that other than this lack the human is a typical human; the 'feelingless' human is aware and they act and think/reason in the same manner as all typical humans. Is the 'feelingless' human devoid of a mind? The denial of a mind to this human would be hard to reasonably defend; it is seemingly obvious that they have a mind. This implies that feeling states are *not necessary* for the existence of a mind. So, if one wishes to draw a *conceptual distinction* in the universe between the 'minded' and the 'non-minded', then one should not make it between those parts of the universe which contain 'feeling states' and those which do not. Feeling states having nothing to do with 'having a mind'; that is to say, feeling states are not a mental attribute.

4.2.4 Perception?

Men have minds, that is to say, they perceive

(Armstrong, 1981, p.1)

mind a vague term, covering...perception

(Gregory, 2004, p.xviii)

As the above claims from Armstrong and Gregory make clear, some humans believe that perceiving is a 'mental' attribute. If this is correct then it means that a part of the universe which perceives 'has a mind'; if it is not correct, then perception is not a 'mental' attribute. It is plausible to claim that the parts of the universe that have a mind *also* have states of perception; however, to say that a part of the universe has a mind *because* it contains states of perception – that perception is a 'mental' attribute – is a very different claim. After all, according to 'pansensism' and 'panperceptualism' – which we will encounter in *Section 5.2* – the entire universe is pervaded with states of perception, but only certain parts of the universe are minded. So, to say that the *'parts of the universe that have a mind also have states of perception'*, is clearly *not* to say that there is any link whatsoever between perception and mind.

Does one really want to assert that if a state of perception exists that this entails that a mind exists? To illuminate the point – it is intelligible that atoms might perceive their surroundings but be wholly devoid of thought, feeling, freedom and awareness; if one asserts that perception is a 'mental attribute' then one is asserting that these atoms have minds. I don't really think that one would want to assert that every atom has 'a mind' simply because atoms have perceptions of their surroundings.

So, the question becomes: Is it intelligible to suppose that atoms can perceive their surroundings? The answer that one gives to this question will probably depend on one's view of the nature of perception. One can either have a 'thin' or a 'thick' view of perception. According to the 'thick' view a state of perception is necessarily accompanied by other attributes; one might believe that perception is necessarily accompanied by awareness, or one might believe that perception necessarily entails the possession of concepts. According to the 'thin' view there can be perception without awareness *and* perception is wholly devoid of concepts. If one has a 'thin' view of perception then one is much more likely to believe that atoms can perceive. If one has a 'thick' view of perception then believing this would entail believing that atoms have concepts or awareness. So, on the 'thin' view of perception it is intelligible to believe that atoms perceive, and such a belief does not entail that atoms have a mind.

Recall that in *Section 3.2* I defended the idea that perception is not concept-entailing because concepts only arise when perceptions of the 'blobject' are thought about. In *Section 6.4.1* I will consider the evidence for perception without awareness and defend the idea that this occurs. So, we have good reasons for believing the 'thin' view of perception to be correct. Perception does not entail awareness or concepts – perception is not 'mind-entailing'.

Finally, let us consider whether a mind can exist without perception. Imagine a human that is wholly devoid of states of perception, but that is still aware and is still engaged in a high level of thinking/reasoning; a level analogous to a typical human (imagine a human lying in a hospital bed who is still able to think in this way but who has lost their ability to perceive). Would you want to assert that this 'perceptionless' human lacked a mind? I assume that the overwhelming majority of humans would assert that the 'perceptionless' human has a mind. If this is right, then states of perception are neither necessary nor sufficient for the existence of a mind; these states are not mental attributes.

4.2.5 Intentionality?

> Philosophers have long been concerned with the phenomenon of intentionality, which has seemed to many to be a fundamental feature of mental states
>
> (Gregory, 1987, p.383)

The phenomenon of intentionality – of aboutness – is often envisioned as a fundamental mark of a mind. Franz Brentano claimed that it is intentionality that creates a division in the universe between the 'mental' and the 'non-mental' because all and only mental phenomena exhibit intentionality. The aim of this section is to explore and reject this idea.

What exactly is a state of intentionality? At first glance it seems to be a rather vague thing which has the potential to encapsulate a diverse range of states. The vagueness of the bounds of intentionality starts with phenomena such as tree rings: Are tree rings intentional states which are 'about' the age of a tree? Perhaps they are not because they are instances of *derived intentionality* rather than *original intentionality*; derived intentionality being a state of intentionality that requires a perceiver in order to exist. So, it is seemingly *original intentionality* that has the potential to divide the universe into the mental and the non-mental.

What is *original intentionality*? States of perception are instances of *original intentionality*; they are states which are *directed towards* that which is perceived. However, as we saw in *Section 4.2.4*, to say that states of perception exist is *not* to say that a mind exists. So, this means that we now have to conclude that instances of *original intentionality* can exist in the absence of a mind. In other words, intentionality cannot divide the universe into the mental and the non-mental.

It is seemingly the case that thinking entails intentionality, and it is very plausible that thinking is necessary and sufficient for the existence of a mind. However, if this is so, then it is thinking which creates a division in the universe between the minded and the non-minded; intentionality is a 'red herring' – not the mark of anything. The part of the universe that contains states of intentionality could be far greater than the part that is minded.

This view – that intentionality is a 'red herring' – chimes with the views of philosophers who advocate the 'physical intentionality' thesis. According to this view all chemical and physical interactions have intentionality because they are directed towards particular outcomes. For example, George Molnar (2003, p.61) states:

> I think that the Brentano thesis is basically mistaken…I accept the intentionality of the mental, and go on to argue that something *very much like* intentionality is a pervasive and ineliminable feature of the physical world.

This belief – that intentionality is more pervasive in the universe than the minded – has inevitably caused many philosophers, including Molnar, to conclude that intentionality cannot be sufficient for the existence of a mind. Of course, another option is to embrace 'physical intentionality' and to argue that this means that minds pervade all of the physical. So, if the 'physical intentionality' thesis is true this means either that intentionality is not a mark of a mind or that minds pervade all of the physical. If one accepts the 'physical intentionality' thesis it is unlikely that one will believe that the existence of a state of intentionality is sufficient for the existence of a mind; this means that the possibility of physical intentionality should multiply our pre-existing doubts that intentionality is sufficient for the existence of a mind.

When you reflect on the question of why you believe that at least most humans possess a mind, I suspect that you are most unlikely to conclude that *it is because humans contain states that are directed towards other phenomena*. It is more likely that you'll conclude that most humans possess a mind because they have awareness or because they think. A state of intentionality by itself is unlikely to cause one to assert – *oh yes, this is what a mind is!* Intentionality might be necessary for the existence of a mind; however it is not sufficient for a mind to exist and is therefore not a mental attribute – parts of one's surroundings can contain states of intentionality without containing a mind.

4.2.6 Awareness?

> **mind** a vague term, covering...awareness
>
> (Gregory, 2004, p. xviii)

It is often claimed that there is a link between awareness and mind; in the above passage Gregory claims that mind is a term 'covering' awareness. The aim of this section is to explore and reject the idea that such a link exists. In *Section 4.1* the *seemingly obvious fact* that humans are parts of the universe that oscillate into and out of awareness was outlined. The question before us now is whether this oscillatory process is related to which parts of the universe have minds.

Let us consider a thought experiment – the 'coma' scenario. In this scenario there is a human that has permanently lost the capacity for awareness, yet they still have a continuously thinking brain. The question is whether the human in the 'coma' scenario possesses a mind. In other words: Is thinking sufficient for the existence of a mind? The most reasonable answer to this question, I suggest, is that thinking is sufficient for the existence of a mind irrespective of whether or not it is accompanied by awareness, and irrespective of whether or not the capacity exists for there to be awareness of this thinking. If this is right, then the reason that a typical human has a mind is that they think.

As a contrast to the phenomenon of mind without awareness it is worth considering the possibility that a human could have both a mind and awareness continuously. In other words, due to some kind of 'malfunction' a human stops oscillating, but rather than entering the 'coma' state they become permanently stuck in the 'on' state. This possibility has supposedly been realised in a few humans such as Thai Ngoc who claims to have been in a continuous state of awareness for thirty-three years (Ngoc, 2010). It is hard to know how seriously one should take such claims but there are usually exceptions to the rule, and there is no good reason to rule out the possibility that a human could become stuck in the 'on' state. In such a scenario, although awareness and thought have become perfectly positively correlated, the reason that a mind exists is due to the existence of thought, not awareness.

These various scenarios imply that there are several possible relationships between mind and awareness, and suggest that whilst the two phenomena may co-occur that they are fundamentally different phenomena. How might one object to this conclusion? One could insist that without awareness, or the capacity for awareness, that there cannot be a mind. This means that in the 'coma' scenario one has three options. One could reject the assumption that thinking is sufficient for a mind; one could reject the assumption that there can be thinking without awareness; or one could reject the assumption that the capacity for awareness has been permanently lost. Given the plausibility of the first two assumptions, the most plausible objection is to assert that the capacity for awareness remains.

However, whilst this objection is the most plausible one it is also deeply problematic. Let us consider a situation in which a human oscillates from an 'in' state to an 'out' state' and then back to an 'in' state. It would be quite right to assert that when this human was not aware that they contained the capacity for awareness. Contrarily, in the 'coma' scenario the arrangement of the universe which provides the capacity for awareness in typical humans does not exist. So, to say that the capacity for awareness still exists in the 'coma' scenario is *really* to say that an arrangement of the universe exists which if it was arranged slightly differently would have the capacity for awareness. In other words, the human in the 'coma' scenario has undergone a change (seemingly in the arrangement of their brain) which means that they are continuously devoid of awareness, and *if* this change was reversed *then* they would be able to oscillate once more. The problem with this is that it is very difficult to defend the idea of a 'slightly' different arrangement. If it is acceptable to assert that an arrangement of the universe which would have the capacity for awareness if it was 'slightly' different is an arrangement of the universe which has the capacity for awareness, then why shouldn't one be able to assert that an arrangement of the universe which is 'slightly more' different also has this capacity. This leads straight to a *reductio ad absurdum* in which arrangements of the universe such as a mound of sand have the capacity for awareness. Either a capacity exists or it doesn't and in the 'coma' scenario it doesn't.

Having explored the possible existence of a mind in the absence of awareness, let us turn to the possibility that states of awareness could exist in the absence of a mind. There seems to be no reason to rule out this possibility. After all, it is intelligible to suppose that a state could exist which is a state of 'simple awareness'. The human brain is a complex arrangement of the universe, but the part of it that is involved in awareness might not be that complex – generating awareness might actually turn out to be relatively simple *if* one knew how to do it. It is possible that awareness could simply be a particular vibrational state which could potentially exist in many non-brain parts of the universe.[40] So, imagine that a magnet contains this vibrational state; as the magnet becomes attracted to particular bits of metal in its surroundings it becomes aware of these attractions. If this were so, I don't think that we should conclude that a magnet has a mind; we should conclude that there is awareness but no mind. Of course, one *could* believe *anything*. The issue at stake is what it is reasonable to believe. One presumably believes that a typical human has a mind; this is a very reasonable thing to believe. One could believe that a magnet has a mind simply because it contains a vibrational state which causes it to have momentary flashes of awareness; however, I take it that the overwhelming majority of humans would believe that

---

[40] I will defend this possibility in *Section 6.9.*

157

this is not a reasonable thing to believe. And what is reasonable is all we have to go on.

What is our conclusion? It is possible that a mind can exist in the absence of awareness so awareness is not necessary for the existence of a mind. It is also possible that a mind can exist in the absence of the capacity for awareness so the capacity for awareness is not necessary for the existence of a mind. Furthermore, it is also possible for awareness to exist in the absence of a mind so awareness is not sufficient for the existence of a mind. So, awareness is not a mental attribute.

4.2.7 *The Inner Cause of Certain Movements in Suitable Circumstances?*

> perhaps what we mean by a mental state is some state of the person that, under suitable circumstances, *brings about* a certain range of behaviour. Perhaps mind can be defined not as behaviour, but rather as the inner *cause* of behaviour... I believe that this is the true account, or, at any rate, a true first account, of what we mean by a mental state.
>
> (Armstrong, 1981, p.7)

According to Armstrong having a 'mind' is having *inner states which under suitable circumstances cause certain types of movements*. This will probably strike one as an extremely odd definition of 'a mind'! It is certainly a bit of a mouthful due to the inclusion of the qualifications. This contorted definition seems to have arisen because of Armstrong's desire to connect having a mind with movement/behaviour.

So, the starting assumption is that certain types of movements mean that 'a mind' exists. All humans that are alive move; even humans that are totally paralysed move when they are moved by other humans. Surely these movements do not mean that the totally paralysed human has a mind! This is why Armstrong states that the movements must arise from "inner states" rather than external states. Some of the movements of humans are involuntary tics and spasms, and automatic responses to external states. Surely these movements are not what having 'a mind' is about! This is why Armstrong says that only certain movements entail that 'a mind' exists. Then we come to "suitable circumstances". So, the "inner states" ability to cause movement can be blocked if the circumstances are not right. So, if a human is paralysed then they might have these "inner states" but not the circumstances suitable for moving. For Armstrong these "inner states" are clearly central to having 'a mind'; he wants to assert that if these states exist that 'a mind' exists *even if* the circumstances are not suitable for movement.

What are these "inner states"? They are 'thinking' states; some of the movements of humans are caused by them pondering/thinking/rationalising. So, why complicate the situation? Why not simply say that to think/ponder/rationalise is to have 'a mind'? A simple definition – to have a mind is to think. A bizarre definition – *to have a mind is to have inner states which under suitable circumstances cause certain types of movements*. Clearly, if one *starts* from a 'mind=cause movement' hypothesis then in order to make this conceptualisation of mind remotely plausible one has to weigh down the definition with qualifications – 'inner', 'suitable', 'certain'. If you have to add this many qualifications you should give up and go back to the 'drawing board'.

Let us further consider the phenomena of 'mind' and 'movement'. That there is no connection between having a mind and movement has been argued for by Galen Strawson (1994, p.251) who states that:

> The Weather Watchers are a race of sentient, intelligent creatures. They are distributed about the surface of the planet, rooted to the ground, profoundly interested in the local weather. They have sensations, thoughts, emotions, beliefs, desires. They possess a conception of an objective, spatial world. But they are constitutionally incapable of any sort of behaviour, as this is ordinarily understood. They lack the necessary physiology. Their mental lives have no other-observable effects. They are not even disposed to behave in any way.

Strawson is surely right to claim that despite the lack of "other-observable effects" the Weather Watchers possess minds. It is very plausible to assert that the mind of a human is the inner cause of certain movements in suitable circumstances. However, this doesn't mean that *what it is to be a mind* is to be the inner cause of certain movements in suitable circumstances. Rather, *what it is to be a mind* has nothing essentially to do with movement – to have a mind is to think.[41]

---

[41] You might find the claim that what it is to be 'a mind' has 'nothing essentially to do with movement' to be too strong a claim – after all it has been accepted that minds can cause movement. It might help to consider what it is to be 'a car'.

4.2.8 Thinking?

> Reason, Descartes says, is a 'universal instrument'... That is, it does not respond specifically to specific stimuli, but varyingly to varying stimuli.
>
> (Matthews, 2005, p.73)

I have claimed that to have a mind is to think. What this means is that if part of the universe is thinking then this part of the universe is a mind. More than this, to say that a mind exists is to say nothing more than to say that thinking exists. The idea that a mind could exist in the absence of thought would be a contradiction in terms. Thinking is both necessary and sufficient for the existence of a mind. Whilst this seems obviously true (I hope you agree) the exact nature of thinking is far from clear. Stephen Priest (1991, p.213) claims that:

> Thinking may take place in language, in an ordinary language such as English, or in an artificial language such as logical notation. Some thinking also takes place in neither of those media but in mental images

---

Cars are designed to move, and are often seen moving, but what it is to be 'a car' seemingly has nothing to do with movement. If one goes to a car park and sees hundreds of non-moving objects that have bonnets, wheels, tyres, engines, steering wheels, etc., it would be acceptable to believe that *these objects are cars*. This is the case even if these cars never move – non-moving cars are cars!

Thinking is clearly a complex phenomenon. Can we have a simple definition of thinking? An adequate initial definition of thinking is that *thinking is a process of reasoning*. Another way of putting this is to say that *thinking is the consideration of possibilities*.[42] Let us connect this definition of thinking=mind to some of the themes of the previous sections in this chapter. Given this definition thinking will entail intentionality – certain things will be reasoned, certain possibilities will be considered (*Section 4.2.5*). However, thinking can clearly exist without any movement or "other-observable effects" (*Section 4.2.7*). Thinking can also exist without awareness (*Section 4.2.6*), without perception (*Section 4.2.4*) and without freedom (*Section 4.2.2*). Finally, thinking is most definitely not a 'central core' of disparate attributes (*Section 4.2.1*).

A distinction is often made between 'conscious thought' and 'non-conscious thought'. I should say a few words about this distinction. A process of reasoning can occur in the absence of the awareness of this process; when such a process occurs it is a process of 'non-conscious thought'. There can also be awareness of a process of reasoning; when there is such awareness the process is often referred to as 'conscious thought'. The phrase 'conscious thought' can conjure up in the imagination visions of something very different from 'non-conscious thought'. There is no great difference but there is a difference. 'Non-conscious thought' has a range of possible inputs which determine which

---

[42] On this definition a chess computer would seemingly be thinking when it is 'considering' the possible moves it can make. Perhaps the word 'considering' is not appropriate in this case. Perhaps 'considering' should be taken in the Cartesian sense – 'considering'/'reasoning' is a 'universal instrument' – it does not respond specifically to specific stimuli, but varyingly to varying stimuli.

thought arises in the next moment: the previous thoughts, perceptions (without awareness), 'feeling states' and memories. 'Conscious thought' has one more possible input which is missing from 'non-conscious thought' – awareness itself can be an input which plays a part in determining which thought arises next in the process.

There is a lot to be said for simplicity – to have a mind is simply to think – and to think is to reason, to consider possibilities. According to this view, none of the other attributes that we have considered – freedom, feeling states, perception, intentionality, awareness, and the inner cause of certain movements in suitable circumstances – are mental attributes.

1.3 Conclusion

In this chapter we have considered 'thick' and 'thin' conceptualisations of both consciousness and mind. I have supported the 'thin' conceptualisations of both phenomena. To have a mind is simply to think; to be conscious is simply to be aware.[43] However, 'consciousness' is an unhelpful term which I will seek to use as sparingly as is possible. It is a term which is frequently used in accordance with the 'thick' conceptualisation to refer to states of 'what-it-is-likeness'=awareness. On the 'thin' conceptualisation of consciousness, 'what-it-is-likeness' and 'awareness' are two distinct phenomena; so the term 'consciousness' could be used to refer *either* to states of 'what-it-is-likeness' *or* to states of awareness. This is why the term is best

---

[43] There is no necessary relation between these two phenomena – thinking and awareness. It is possible that a part of the universe can think without ever being aware, and that a part of the universe can be aware without ever thinking.

discarded; the only terms we need are 'awareness' and 'what-it-is-likeness'.[44]

If one believes that very few parts of the universe (humans and possibly a few non-human parts too) are conscious and/or minded, and adopts a 'thick' conceptualisation of consciousness and mind, then one will inevitably believe that there is a great chasm between humans and the overwhelming majority of their surroundings. However, if one adopts a 'thin' conceptualisation of these phenomena then one will either believe that there is a much smaller chasm or that there is no chasm.

I have tried to convince you that there are very good reasons for adopting a 'thin' conceptualisation of both consciousness and mind. The 'thin' conceptualisation of consciousness largely demystifies the "problem of consciousness"; this is clearly preferable to the 'thick' conceptualisation which entails a seemingly intractable mystery. The 'thin' conceptualisation of mind is preferable to the 'thick' conceptualisation because when we think seriously about what 'having a mind' entails we are left with the conclusion that to have a mind is simply to think. All of the other various attributes which are often called 'mental' attributes – and which jointly form the 'thick' conceptualisation of mind – are actually 'non-mental' attributes.

I have reached the conclusion that 'having a mind' is simply to think without taking the evolutionary perspective.

---

[44] I will occasionally refer to the 'thin conceptualisation of consciousness' rather than to 'awareness'. When I use this term I will be referring solely to states of awareness, and *not* to states of 'what-it-is-likeness'. Any use of the term 'consciousness' should refer to states of awareness – to be 'conscious of' is to be 'aware of'.

When we take this perspective we have an additional reason to reach this conclusion. Let us presume that in the immediate aftermath of the Big Bang that the universe did not contain any minds; there must have been an exact moment when the first mind came into existence. Was this moment the coming into existence of a single attribute? Or, was this moment the simultaneous coming into existence of multiple attributes? It seems more rationally palatable to believe that when the first mind evolved that this event entailed the coming into existence of a single attribute. If this is right, then it means that attributes such as 'feeling states', perception and intentionality could have existed in the evolving universe before there were any minds; these attributes would be non-mental attributes. The moment that a single attribute evolved – thought – would have been the moment that the first mind came into existence.

Chapter 5

'What-it-is-likeness' and the human senses

In *Chapter Four* we considered the question of how similar humans are to the surroundings which evolved them from the perspective of the conceptualisations that humans make of the phenomena of consciousness and mind. In this chapter our focus switches to two specific human attributes which are often thought to carve a great division in the universe between humans and their surroundings – 'what-it-is-likeness' and the human senses.

The purpose of this chapter is to explore two issues. Firstly: what exactly is 'what-it-is-likeness'? Which states of the universe are encompassed within this term? And, which of these states does the 'panwhat-it-is-likeness' advocate believe to be pervasive in the universe? Secondly: how many senses does a typical human have? And, given that humans evolved from their surroundings, how many of the five 'traditional' human senses could be pervasive in the universe? Clearly, if a large part of the family of 'what-it-is-likeness' states pervade human surroundings, and if most of the five 'traditional' human senses pervade human surroundings, then humans will be very similar to the surroundings which evolved them. Contrarily, if human surroundings are largely, or wholly, devoid of these phenomena, then humans are seemingly very different to the surroundings which evolved them.

In *Section 5.1 – What is 'what-it-is-likeness'?* – I claim that the family of 'what-it-is-likeness' states contains states of feeling such as those that a human might refer to as pain, tingling, pins and needles, dread, jealousy, love and fear, and also tasting, smelling, touching, seeing and hearing. I then consider all of the 'what-it-is-likeness' states *except* the 'what-it-is-likeness' of the five traditional senses – tasting, smelling, touching, seeing and hearing. I contend that *in terms of their 'what-it-is-likeness'* all of these states are types of 'feeling states'. That is to say, all of these states are part of a homogenous group of states – 'feeling states' – that are fundamentally of the same type.

In *Section 5.2 – How many human senses are there?* – I turn to the issue of the human senses. I consider the question of how many human senses there are from the perspective of the phenomenon of 'what-it-is-likeness'. I suggest that the typical human possesses only two senses – seeing and hearing, and that the other three traditional 'senses' are actually parts of the family of 'feeling states'. This means that there are two very distinct groups of 'what-it-is-likeness' – 'feeling states' and seeing/hearing 'sensory' states.

In *Section 5.3 – The location of 'feeling states'* – I consider the question of where the 'feeling states' that a human becomes aware of are located. This is an important issue because if these 'feeling states' are solely located in human brains then this supports the idea that the non-brain universe is wholly devoid of 'feeling states'. However, if these states pervade the human body then this supports the idea that 'feeling states' pervade the entire non-brain universe.

In the following two sections we turn to the 'what-it-is-likeness' of the two human senses. In *Section 5.4 – The location of the 'what-it-is-likeness' of the visual sense* – I consider the question of where the 'what-it-is-likeness' of the visual sense is located, and in the following section – *Section 5.5* – I do the same for the auditory sense. Finally, in *Section 5.6*, I draw some conclusions.

5.1 What is 'what-it-is-likeness'?

> There is no clear reason why it should feel like anything to be alive.
>
> (Robert Wright, cited in Barlow, C., 1991, p.246)

One can conceive of a universe that is wholly devoid of 'what-it-is-likeness' – in this universe it would *not* "feel like anything to be alive"; the very fact that it *does* feel like something to be alive in this universe is a fact that needs to be pondered. When one talks of 'what-it-is-likeness' one is, in large part, referring to the fact that in this universe *it does* feel like something to be alive. We have already encountered two 'world views' – panwhat-it-is-likeness and panawareness – according to which, in the universe in which we live, it feels like something to *not* be alive; on these views it feels like something to be a stone and a hydrogen atom.[45] The panwhat-it-is-likeness advocate and the panaware-ness advocate have a compelling solution to Wright's musings; on these views the reason that it feels like something to be alive is that the entire universe is pervaded with feeling states.

---

[45] Recall that panwhat-it-is-likeness entails that all parts of the universe contain 'feeling states' but that very few states of the universe contain awareness. Panawareness entails that all parts of the universe contain 'feeling states'=awareness.

What exactly does it mean to say that it *feels like something* to be alive? Primarily, what it means is that one becomes aware of a range of feelings which one may find to be very pleasant, very unpleasant, or somewhere in-between. When one attempts to communicate these feelings one might use words such as 'throbbing pain', 'headache', 'tingling', or 'pins-and-needles' to refer to them. Let us use the term 'bodily sensations' to refer to this group of 'what-it-is-likeness' states.

A second group of 'what-it-is-likeness' states contains the states of feeling which one might describe as 'dread', 'jealousy', 'nervousness', 'love', 'fear' or 'rage'. How are these states distinguished from the group of 'bodily sensations'? The simple answer is that these states are more complex as they involve either thought or sensory perception. So, if one is kidnapped and one's captors put one – blindfolded and with ears muffled – into a car, one is likely to contain a state which one might describe as 'fear'; this state is a hybrid state – it not only feels a certain way, it also contains a range of *thoughts* such as the pondering of what is going to happen to one. Similarly, if one is about to give a speech to thousands of humans it is likely that one will contain a state which one might describe as 'nervous-ness' – this state does not just feel a certain way, it also contains thoughts and/or perceptions; it can contain the *thought* that one hopes that one speaks clearly, and the *perception* of the sound of the audience. Similarly, if one sees an angelic woman across the room it is possible that one will contain a state which one might describe as 'love' – this state does not just feel a certain way, it also contains the *perception* of the angelic woman. Let us use the term 'emotions' to refer this group of 'what-it-is-likeness' states.

So, if as one is walking down the street a brick which one was totally unaware of falls onto one's head the feeling state which exists as the brick hits one's head is a 'bodily sensation'. However, if as one was walking down the street one saw a brick falling towards one then this perception would both cause a thought within one and feelings of 'what-it-is-likeness' to arise within one. This state, as it contains perception, thought, and 'what-it-is-likeness', is an 'emotion'.

A third group of 'what-it-is-likeness' states contains the 'what-it-is-likeness' of the five traditional senses – the phenomena of tasting, smelling, seeing, hearing and touching. There is an instance of tactual 'what-it-is-likeness' when a finger touches velvet; an instance of auditory 'what-it-is-likeness' when a ringing telephone is heard; an instance of olfactory 'what-it-is-likeness' when burnt toast is smelt; an instance of gustatory 'what-it-is-likeness' when a hot chilli is tasted; and an instance of visual 'what-it-is-likeness' when a red book is seen.

What exactly does it mean to say that there is a 'what-it-is-likeness' to see a red book? In considering the existence of the 'what-it-is-likeness' of seeing, and the fact that some deny that such a thing exists, Galen Strawson (2008, p. 102) states:

> Most, however, will agree that the notion of the qualitative character of colour-experience can reasonably be taken for granted. And for present purposes, a sufficient reply to those who disagree is simply as follows. Consider your present visual experience. Look at the bookshelf. (Get out some of the brightest books.) There you have it.

Are there any more groups of 'what-it-is-likeness'? This is debatable. Some humans believe that there is a 'what-it-is-likeness' to think; according to this view there is something it is like to think: *two plus two equals four*. John R. Searle (2002, p.40) sums up the situation as follows: "Some people think that qualia are characteristics of only perceptual experiences, such as seeing colours and having sensations such as pains, but that there is no qualitative character to thinking." An advocate of the view that there is a 'what-it-is-likeness' to think is Strawson (2008, p.262) who states:

> consider (experience) the difference, for you, between my saying 'I'm reading *War and Peace'* and my saying 'barath abalori trafalon'. In both cases you experience sounds, but in the first case you experience something more: you have an understanding-experience, a cognitive experience. Why isn't this point universally acknowledged? *Have you had only sensory experience for the last two minutes?*

In defence of the view that there is no 'what-it-is-likeness' involved in thinking David Rosenthal (2005, p.10) claims, with reference to his Higher Order Thought theory, that: "HOTs, like thoughts in general, have no qualitative character themselves". Another human with such a view is William S. Robinson (1988, p.86) who asserts that: "thinkings do not have qualia. What thinkings are about may have them." Robinson (1988, pp.86-7) argues that:

> some thinkings may cause feelings, as my believing that Mary is with John may cause jealousy; but a different

171

believing could have caused the same feeling (had I loved Ann rather than Mary) and the same believing could have caused a different feeling (had I never loved Mary). Thus, the feeling quality that may occur together with a believing cannot be held to be a quality of the believing itself.

So, as with 'consciousness' and 'mind', there are both 'thick' and 'thin' conceptualisations of thinking. According to the 'thin' view thinking is simply a process of reasoning that is not 'what-it-is-likeness' entailing. According to the 'thick' view thinking is a process that entails both reasoning and 'what-it-is-likeness'. As with both 'consciousness' and 'mind', I believe that there are good grounds for adopting a 'thin' conceptualisation, and thereby agreeing with Rosenthal and Robinson that thoughts do not entail 'what-it-is-likeness'. These grounds are my own thoughts; however, you might disagree if you have a different kind of thoughts to me. For me, when I think *"two plus two equals four"* this is barely comparable to being in pain, to tasting a chilli, and to seeing some very bright books. These latter events involve 'what-it-is-likeness' but thinking does not. One can have this view and still agree with Strawson that there is a difference between understanding words and not understanding words; contra-Strawson one simply believes that this difference is not a difference in 'what-it-is-likeness'.

If this is right, and I will assume that it is right, then the family of 'what-it-is-likeness' states contains three groups: 'bodily sensations', 'emotions', and the 'what-it-is-likeness' of the five traditional senses – tasting, smelling, seeing, hearing and touching.

5.1.1   The family of 'feeling states'

In this section our focus is on two of the three groups of 'what-it-is-likeness' – 'bodily sensations' and 'emotions'. In particular, my aim is to explore and defend the idea that *in terms of their 'what-it-is-likeness'* there is no difference between these two groups. In the previous section I suggested that 'emotions' are distinguished from 'bodily sensations' because they are more complex – they contain thought and/or perception in addition to feeling a certain way. Now, the 'what-it-is-likeness' of perception is obviously not itself part of the 'what-it-is-likeness' of an 'emotion'; that is to say, the 'what-it-is-likeness' of perceiving the redness of the dress of the angelic woman is not part of the 'what-it-is-likeness' of the 'emotion' of 'love'. Furthermore, given that thought is wholly devoid of 'what-it-is-likeness', thought cannot create a distinction in 'what-it-is-likeness' between 'bodily sensations' and 'emotions'. Given this, it seems to straightforwardly follow that 'emotions' are not distinguished from 'bodily sensations' by differing in respect of 'what-it-is-likeness'. A 'bodily sensation' is simply a state that *feels a certain way*, and the 'what-it-is-likeness' component of an 'emotion' is also a state that *feels a certain way*.

This means that to say that there is a 'feeling of pain' is to say the same kind of thing as to say that there is a 'feeling of dread', a 'feeling of pins and needles', a 'feeling of love', and a 'feeling of calm'. There is a homogenous group that contains the 'what-it-is-likeness' of 'bodily sensations' and the 'what-it-is-likeness' of 'emotions'; exactly the same state of 'what-it-is-likeness' that exists in an 'emotion' could exist in a 'bodily sensation'. It is the addition of a thought and/or a perception to the 'what-it-is-likeness' of a 'bodily sensation' that creates an 'emotion'.

Is there *any* difference in the 'what-it-is-likeness' which exists in 'bodily sensations' and 'emotions'? I have suggested that the answer to this question is no because the same 'what-it-is-likeness' can be present in both a 'bodily sensation' and an 'emotion'. Furthermore, even if the 'what-it-is-likeness' that is part of an 'emotion' could not exist in the absence of the 'emotion' – if only a very similar, but not identical, state of 'what-it-is-likeness' could exist as a 'bodily sensation' – then this close similarity would itself be sufficient to support the idea that the two instances of 'what-it-is-likeness' are part of a homogenous group. That is to say, there is no reason to suppose that there is any general difference in the *level or nature* of 'what-it-is-likeness' which exists in the two groups. There is no reason to suppose that in general 'emotions' entail a higher level (or a lower level) of feeling than 'bodily sensations'. There is also no reason to suppose that in general 'emotions' have a longer duration (or a shorter duration) than 'bodily sensations'.

Let us be clear about what has been claimed. Our conclusion is that the states of 'what-it-is-likeness' that exist within humans as 'bodily sensations' and 'emotions' are part of a homogenous group. Let us refer to this group as the *family of 'feeling states'*. There is a distinction between 'bodily sensations' and 'emotions' but this is not a difference in 'what-it-is-likeness'; the distinction arises because an 'emotion' is a hybrid state which contains thought and/or perception in addition to 'what-it-is-likeness'. It is likely that in many cases there will be a complex relationship between the different components of an 'emotion'; these relations will include causation in different directions and feedback loops. In view of this let us consider the following claim by Peter Goldie (2002, pp.12-13):

An emotion – for example, John's being angry...is complex in that it will typically involve many different elements: it involves episodes of emotional experience, including perceptions, thoughts, and feelings of various kinds, and bodily changes of various kinds; and it involves dispositions, including dispositions to experience further emotional episodes, to have further thoughts and feelings, and to behave in certain ways.

This statement nicely sums up the different components of an 'emotion'. It is clearly concordant with the view being defended in this section according to which there is no difference in 'what-it-is-likeness' between 'emotions' and 'bodily sensations'. An 'emotion' is simply a hybrid state of the world whose 'what-it-is-likeness' component is a 'bodily sensation'. So, there is a difference between an 'emotion' and a 'bodily sensation'. However, the 'what-it-is-likeness' of all instances of 'emotions' *and* 'bodily sensations' is part of one homogenous group – the family of 'feeling states'.

5.2 *How many human senses are there?*

In the last section we saw that what we initially took to be two different groups of 'what-it-is-likeness' – 'bodily sensations' and 'emotions' – are actually a single group of 'what-it-is-likeness': 'feeling states'. It is time to turn our attention to the remaining group of 'what-it-is-likeness' states – the 'what-it-is-likeness' of the five traditional senses. Our consideration of this group of 'what-it-is-likeness' states starts by considering the question of how many human senses there are. It will be suggested that there is a sharp disparity within the group of five traditional senses – two senses have one type of 'what-it-is-likeness', and the other three senses have a different type of 'what-it-is-likeness'. Furthermore, from the evolutionary perspective, it will be suggested that this difference in 'what-it-is-likeness' is potentially of great significance. This is because it is possible that the group of three senses has not evolved, whilst the group of two senses has evolved. The question of what a 'sense' is is then addressed and it is suggested that, in accordance with the previous divisions, a typical human has only two senses; what one refers to as their other three senses are actually parts of the family of 'feeling states'.

The traditional idea that there are five different human senses has a long history. In *De Anima* Aristotle gives a detailed analysis of the five senses and concludes that: "there is no sixth sense in addition to the five enumerated – sight, hearing, smell, taste, touch" (2001, p.581). According to Matthew Nudds (2009) this distinction between five different human senses has pervaded human history as it "is not a recently invented distinction nor an artefact of our scientific culture, but has been made for at least as long as people have been able to record the fact".

The main challenge to the traditional idea has come from those who claim that there are more than five human senses. For example, from the scientific perspective it has frequently been claimed that there is a vestibular sense which detects balance and a proprioceptive sense which detects the orientation of one's limbs in space. Furthermore, it has also frequently been claimed that there is a 'sixth sense' which detects danger and provides a 'gut instinct'.

What are we to make of the idea that a human could have more than five senses? It might seem fairly harmless to assert that one has a proprioceptive 'sense' because one has an idea of the orientation of one's limbs in space, and that one has a vestibular 'sense' because one has an awareness of whether one is in a state of equilibrium. However, this extension of the concept of a 'sense' is unhelpful; if the extension was continued the conclusion could be reached that there are hundreds of human senses.[46] We surely do not want to go down this path. Let us assume that Aristotle was correct when he argued that there are not more than five human senses. In other words, if one asserts that there is a vestibular 'sense', or a sixth 'sense', then one is actually referring to a non-sense phenomenon.

I have claimed that there are not more than five human senses. The question remains as to whether the traditional idea is correct. Could there actually be less than five human senses? In the rest of this section I will suggest that the traditional idea is false because *prior to conceptualisation* there are actually only two human senses. How will this conclusion be attained? Firstly, I will contend that there is a sharp disparity within the group of five traditional senses – two senses operate in one way and the other three senses operate in a totally different way. Secondly, I will address the question of what a 'sense' is and will claim that only the former group of two senses are senses. Thirdly, I will suggest that the type of 'what-it-is-likeness' possessed by the group of three 'senses' makes these 'non-senses' part of the family of 'feeling states'.

Let us first consider what are traditionally referred to as the 'sense of taste' and the 'sense of smell'. Are these two distinct 'senses'? Carolyn Korsmeyer (2009) claims that: "some researchers group taste and smell together in terms of "mouth-sense"." Clearly, according to these researchers a typical human

---

[46] Later in this section I will address the question of what a 'sense' is.

would have four senses rather than five. So, on this view, post-conceptualisation one asserts that one 'tastes with one's tongue' and 'smells with one's nose', but in reality there is a singular "mouth-sense". This possibility for collapsing two senses into one opens up the way for a deeper reconceptualisation.

What is occurring when one asserts that one 'smells with one's nose'? What one is referring to is a particular interaction between two parts of the universe. For simplicity I will use the word 'particles' to refer to parts of the universe. So, smelling involves particles entering one's nose and interacting with particles in that nose. This interaction involves a particular state of 'what-it-is-likeness' – a state which, when one becomes aware of it, one calls a 'smell'.[47] Similarly, when one asserts that one 'tastes with one's tongue' what is occurring is that non-tongue particles are interacting with the particles of one's tongue. This interaction involves a particular state of 'what-it-is-likeness' – a state which one calls a 'taste'. Let us consider a particular particle – P*. If P* interacts with one's nose then when one becomes aware of this interaction one would assert that one smelt something; but if P* had interacted with one's tongue then one would have asserted that one tasted something. The question is: Are these two different types of interactions fundamentally of the same kind? In other words, *prior to conceptualisation* is the 'what-it-is-likeness' which is involved in smelling of the same kind as that which is involved in tasting?

---

[47] A human who is sitting in their lounge might have the following thought: 'I can smell burnt toast in the kitchen'. There is a 'what-it-is-likeness' to smell burnt toast. This 'what-it-is-likeness' is clearly not located in the kitchen; it is located in the nose of the human. The thought clearly means: *the state of 'what-it-is-likeness' in my nose leads me to believe that there is burnt toast in the kitchen.*

The answer that one gives to this question will be influenced by how one conceives of one's surroundings. If one conceives of one's surroundings as largely devoid of 'what-it-is-likeness' (as the physicalist does[48]) then one will conceive of P* itself – which is a very simple part of the universe – as wholly devoid of 'what-it-is-likeness'. On this view the 'what-it-is-likeness' of smelling emerges – wholly springs into existence – when P* interacts with the particles of one's nose, and the 'what-it-is-likeness' of tasting wholly springs into existence when P* interacts with the particles of one's tongue. Given that a nose and a tongue are very different arrangements of the universe, on this view one is likely to envision that the 'what-it-is-likeness' of smelling is of a very different nature to the 'what-it-is-likeness' of tasting.

If one has a different conceptualisation of one's surroundings then one is likely to reach a different conclusion. So, if one has the 'world view' of panwhat-it-is-likeness then one will have a very different conception of the state of P* prior to its interaction with one's nose and tongue. One would conceive of P* as being a state of 'what-it-is-likeness'; according to this view 'what-it-is-likeness' does not wholly spring into existence when P* interacts with one's nose or tongue. Furthermore, every interaction which P* is involved in is believed to involve 'what-it-is-likeness'. According to this view the 'what-it-is-likeness' involved in the interactions which one calls 'smelling' and 'tasting' is of the same kind as that which exists when two stones collide. In other words, for the panwhat-it-is-likeness advocate, P* itself is a certain 'feeling state' (F*) which is *not divided* into

---

[48] Recall that I am now using the term 'physicalist' to refer to the 'world view' of 'emergent physicalism'.

'smell qualities' and 'taste qualities', but which *is a large part* of the 'what-it-is-likeness' that one asserts is a 'smell' or a 'taste'.[49] So, when P* interacts with the particles of one's nose a new 'feeling state' is formed by the interaction of F* with the 'feeling states' of the particles of one's nose. Similarly, if P* were to interact with the particles of one's tongue a new 'feeling state' would be formed by the interaction of F* with the 'feeling states' of the particles of one's tongue. Similarly, if P* were to interact with particles in a stone a new 'feeling state' would be formed by the interaction of F* with the 'feeling states' of the particles in the stone. For the panwhat-it-is-likeness advocate these events are all of the same kind. So, on this view, it would strictly be correct to say that the 'what-it-is-likeness' that is involved when P* interacts with the particles in a stone contains the 'what-it-is-likeness' of 'taste' and the 'what-it-is-likeness' of 'smell'.

Let me elucidate this a little. When a human asserts that they 'smell' something they are referring to a 'what-it-is-likeness' entailing interaction; this interaction is occurring between a particle that has entered their nose (P*) and particles in their nose. This event is typically conceptualised as a one-way event – there is something 'objective' (something wholly devoid of 'what-it-is-likeness') which is *being smelt* by a human. From the panwhat-it-is-likeness perspective we need to move to thinking in terms of two-way events. The 'what-it-is-likeness' of smell is generated by an interaction of two particles – the particles of the nose are interacting with P*; furthermore, exactly the same types of interactions are occurring throughout

---

[49] According to this view F* is a 'feeling state' and the particles of one's nose and tongue are also 'feeling states' (F^N and F^T). The F*/F^N and F*/F^T interactions bring forth a new 'feeling state' which is constituted by F*/F^N and F*/F^T respectively.

the universe, such as when two stones collide. Another way of putting this is to say that the 'what-it-is-likeness' which occurs when *any* two parts of the universe interact is of the *same* type. It is obviously fine if humans want to create words and refer to particular interactions as 'smells' and 'tastes' – the thing to be realised is that on the panwhat-it-is-likeness view the 'what-it-is-likeness' involved in these 'smells' and 'tastes' is of exactly the same type as that which pervades all of the universe. Noses and tongues do not bring a 'new' type of 'what-it-is-likeness' into existence.

Now, it should be noted that the physicalist could also assert that the 'what-it-is-likeness' of smell is of the same kind as the 'what-it-is-likeness' of taste, whilst still denying that there is 'what-it-is-likeness' involved when two stones collide. However, the denial of 'what-it-is-likeness' to P* makes this view seem highly implausible. This is because the advocate of this view accepts that *all* of the 'what-it-is-likeness' of 'smell' is generated *wholly* by one's nose, and that *all* of the 'what-it-is-likeness' of 'taste' is generated *wholly* by one's tongue; and one's nose is clearly a very different arrangement of particles to one's tongue.

Let us turn to the interactions which are occurring when one asserts that one has a 'sense of touch'. These interactions are analogous to the interactions which we have already encountered with smell and taste. So, touching involves some of the particles of one's skin interacting with other particles. This interaction involves a particular state of 'what-it-is-likeness' at the location of the interaction – a state which one calls a 'touch'. As with smell and taste the physicalist will, in all likelihood, hold that a unique type of 'what-it-is-likeness' is emerging when one touches. Whilst, the panwhat-it-is-likeness advocate will hold that what is occurring when one touches is an event of the same kind as that which occurs when one is smelling or tasting.[50] In other words, when one touches velvet what is happening is that P* (particles in the velvet) are interacting with some of the particles of one's skin. In this event the 'feeling states' of P* (F*) interact with the 'feeling states' of the particles of one's skin to create a new 'feeling state'.

Let us recap. We are considering the possibility that there are two, three or four senses. As the first part of this endeavour we have been concentrating on three of the five traditional senses – taste, smell and touch. In particular, we have been considering the different ways in which the physicalist and the panwhat-it-is-likeness advocate conceive of the 'what-it-is-likeness' interactions which are occurring in the operation of these senses. The physicalist will typically advocate that there are different types of 'what-it-is-likeness' involved in taste, smell and touch. Whilst, for the panwhat-it-is-likeness advocate what post-conceptualisation one refers to as 'tasting', 'smelling' and 'touching' are in fact events of a homogenous type – the same type of 'what-it-is-likeness' is present in *all* of these events.

There is a very important similarity between these three traditional senses – each of them involves 'what-it-is-likeness' at a *particular location* (nose, tongue or skin) when particles interact at that location. From the panwhat-it-is-likeness perspective these what-it-is-likeness interactions pervade the universe. Let me elucidate this a little. According to this view, every time that two parts of the universe come into contact this involves a state of 'what-it-is-likeness' – a 'feeling state' – and

---

[50] By 'the same kind' I mean that these events are not different in any significant way – there are not different types or levels.

this is of the same type as that which exists when a human touches, tastes or smells. Now, one might intuitively believe that there is some kind of distinction between the 'what-it-is-likeness' of smell, taste and touch. However, when one thinks about it one realises that there are a vast array of different states of 'what-it-is-likeness' *within* the realm of taste, *within* the realm of touch, and *within* the realm of smell. There is much more diversity *within* these separate realms than there is *between* certain touches and tastes, tastes and smells, and smells and touches.

So, I am suggesting that when a human has a state of 'what-it-is-likeness' in their nose they call it a smell *because* it is in their nose, if this exact same state of 'what-it-is-likeness' was located in their tongue they would call it a taste, and if it arose after they had touched something they would call this exact same state a touch. All of these 'what-it-is-likeness' states are of the same general kind. According to the panwhat-it-is-likeness view these states pervade human surroundings. The interactions involving human tongues, noses, and skin, are just a tiny fraction of the interactions in the universe; the universe is a massive web of these 'what-it-is-likeness' interactions.

Let us now consider the other two traditional human senses – seeing and hearing. Are the 'what-it-is-likeness' interactions which are occurring when one sees and hears analogous to those which occur when one tastes/smells/touches? I take it that the answer to this question is obviously 'no'. Why is this? It is because seeing and hearing entail the perception of distantly located 'what-it-is-likeness'. I see a red table which is one metre in front of me; I see a white window which is 5 metres in front of me; I see a brown building which is 50 metres in front of me; I see the yellow Sun which is many more metres in front of me; I hear the sound of my watch ticking about 30cm in front of me; I hear a dog barking about 20 metres in front of me; I hear the siren of a police car which is 100 metres in front of me.[51]

You might believe that I claim too much in the last paragraph. It is seemingly obvious that there is a sharp distinction between seeing and hearing and the other three traditional human senses. Seeing and hearing entail the perception of varyingly located distant objects; in contrast, touching/tasting/smelling entails interactions at a single location. One can accept this whilst rejecting the idea that the *'what-it-is-likeness'* of visual and auditory perception is distantly located. I will explore and defend the idea of distantly located visual and auditory 'what-it-is-likeness' in *Sections 5.4* and *5.5*.

Let us now consider what it means to talk of a sense. Are there two groups of senses? Or, is one of these groups actually an impostor – a collection of non-senses? If, pre-conceptualisation, smelling, tasting and touching are actually one phenomenon, and seeing and hearing are two distinct phenomena, then one could hold that there are actually three senses – seeing, hearing, and smelling-tasting-touching. This would be a reasonable position to take.

Another option is to assert that there are only two senses – seeing and hearing. One reason for adopting this approach is that one might find it implausible to assert that the interaction of two particles at a particular location is a sense (one can believe that touching/tasting/smelling is wholly devoid of

---

[51] As we will see in *Sections 5.4* and *5.5* the situation is a little more complex than this implies. To say that I perceive distantly located yellowness (distantly located 'what-it-is-likeness') when I perceive the Sun *is not* to say that 'the Sun is yellow'.

perception/sensing – there are just two particles interacting at a particular location). Seeing and hearing are clearly senses because they involve the sensing of things which are distantly located. In contrast, tasting-touching-smelling are 'what-it-is-likeness' interactions at a particular location. In order to shed some light on this let us consider the tradition known as 'pansensism'; according to this view everything in the universe 'senses'. The most well known pansensists are Bernardino Telesio, Tommaso Campanella and Ernst Mach. According to Campanella (1620 cited in Skrbina, 2005, p.79):

> the world...all its parts and particles thereof [are] to be endowed with sense perception, some more clearly, some more obscurely, to the extent required for the preservation of themselves and of the whole in which they share sensation.

Let us assume that pansensism is true. The 'sensing' ability of particles is most intelligibly thought of as solely entailing 'sensings' at the location of the particle (particles don't sense distantly located 'what-it-is-likeness'). This implies that particle 'sensings' are the same kind of interactions as tasting-touching-smelling. In other words, perhaps tasting-touching-smelling are what-it-is-likeness interactions which exist at a particular location when two particles perceive each other at that location. And, perhaps these interactions pervade the universe In contrast to these interactions the human visual and auditory senses are very different. To mark this difference the former interactions can be thought of as 'particle perceptions' (not senses) and the latter can be thought of as 'sensory perceptions' (senses).[52] If pansensism is true this implies that a human contains an exceedingly large number of 'particle perceptions' and two senses (and 'pansensism' would be better described as 'panperceptualism').

If one believes that 'particle perceptions' *should* be thought of as a sense then one can simply assert that a typical human has three senses. I have already admitted that this is "a reasonable position to take", and not much seems to hang on which terminology one uses. However, if one agrees with me that there is a sharp division between the five traditional human senses then using my favoured terminology in order to reflect this seems to be of some use.

So, rather than hold that simple interactions at a particular location are a 'sense' let us define a sense as *a particular arrangement of the universe which gains access to events which exist at distant locations*.[53] Particles might perceive but they are not senses. On this conceptualisation of a sense a typical human has two senses – an auditory sense and a visual sense.[54] From the evolutionary perspective the panwhat-it-is-likeness advocate will envision the entire universe as being

---

[52] I elucidate this distinction in the next chapter.

[53] The 'particular arrangement' criteria is necessary because it seems that all of the particles in the universe gain access to distantly located particles via quantum entanglement; this access isn't sufficient for the existence of a 'sense'.

[54] For the sake of clarity I should remind you that when one smells one only gains access to the event which is occurring in one's nose; by accessing this event one can make an inference about events at distant locations. If one is in doubt just go outside and smell some flowers – one will find that one becomes aware of a state of 'what-it-is-likeness' that is located in one's nose.

pervaded with three of the traditional five human senses (the 'non-senses'), and the two genuine human senses evolving within humans.

It should be noted that this conceptualisation of a sense entails that arrangements of the universe such as a 'dictaphone', a 'video camera' and a 'mobile phone' are senses; *for these objects are particular arrangements of the universe which gain access to events which exist at distant locations.*[55] This should be seen as a point in favour of the view, because any conceptualisation of a sense which entails that only the parts of the universe which humans call 'living' can contain a sense would be very simplistic and wholly inadequate. Before one reflects on this issue one might just assume that only animals have senses; however, after reflection one will, I believe, conclude that it is obvious that certain non-living arrangements of the universe can also be senses.

I have suggested that smelling-tasting-touching is not a sense; if this is right, then how should we think of these states of 'what-it-is-likeness'? There is a most elegant answer to this question. The states of 'what-it-is-likeness' that exist when one smells, tastes and touches can be thought of as more members of the family of 'feeling states'. So, when one contains states of 'what-it-is-likeness' which one would describe as 'pain', 'taste', 'pins and needles', 'fear', 'touch', 'tingling' and 'smell' – these states are all part of the same homogenous group of states. The members of this group are *all* states where interactions between

---

[55] In defining a sense I have not addressed the issue of how perception is related to awareness; this topic will be covered in *Section 6.4*. For now it is worth noting that to say that a 'video camera' is a 'sense' is *not* to say that a 'video camera' is *aware* of the distantly located events that it is accessing.

particles involve 'what-it-is-likeness' at the location of the interaction.

So, there are two very distinct groups of 'what-it-is-likeness' – 'feeling states' and the 'sensory what-it-is-likeness' of the two senses. The 'what-it-is-likeness' of 'feeling states' is located where the 'feeling state' is located; in contrast, 'sensory what-it-is-likeness' is not located where the sense is located. In the next section our concentration will be on the first group of 'what-it-is-likeness' states – 'feeling states'. Then, in the two subsequent sections, we will consider the 'what-it-is-likeness' of the second group – 'sensory what-it-is-likeness'.

5.3 The location of 'feeling states'

My objective in this section is to consider where the 'feeling states' that a human becomes aware of are located. This is important because the conclusion we reach can either support or undermine the idea that human surroundings are pervaded with 'feeling states' which are analogous to those that a human describes as 'pain', 'pins and needles' and 'tingling'.[56] In the following sections I will consider a number of possibilities concerning the location of the 'feeling states' that a human becomes aware of. After outlining the various positions I then, in *Section 5.3.5*, discuss in further detail why the issue of the location of 'feeling states' is important; finally, in *Section 5.3.6*, I compare the various positions to see which is the most appealing.

---

[56] In *Section 5.2* I claimed that 'touching-tasting-smelling' are also 'feeling states'. In this section I will concentrate on 'feeling states' such as pain and pins and needles; however, the considerations apply to all 'feeling states'.

5.3.1 The location of 'feeling states' solely in the brain (braincentricism)

The dominant philosophical and scientific view is that *the 'feeling states' that a human becomes aware of* are solely located in the brain of that human. According to this view when one has 'feeling states' which one might describe as 'a terrible backache', 'pins and needles in the right leg', 'a stabbing pain in the left foot', 'a tingling feeling all over the body', 'a throbbing in the right thumb' and 'a feeling of serenity', all of these states are solely located in one's brain.

Why would someone (a small number of philosophers) believe that when there is a 'feeling state' that appears to them to most definitely be located in their left foot that this 'feeling state' is actually located in their brain? The short answer is that this is what their 'world view' necessitates *must* be the case; so, scientific findings will be interpreted through this 'world view' and *made to fit* this 'world view'. Advocates of this view of 'feeling state' location often have the 'world view' of physicalism. The physicalist could believe that it is only in certain parts of the universe that 'feeling states' emerge, and that these parts are human bodies; however, in reality the overwhelming majority of physicalists believe that the parts of the universe in which 'feeling states' evolve are brains (not bodies). Indeed, physicalists are typically brain obsessed. For example, John Searle (1992, p.63) asserts:

> Common sense tells us that our pains are located in physical space within our bodies, that for example, a pain in the foot is literally in the physical space of the foot. But we now know that is false. The brain forms a body image,

and pains like all bodily sensations, are parts of the body image. The pain in the foot is literally in the physical space in the brain.

This is a clear example of a human interpreting a scientific finding so as to make it fit their 'world view'. It is worrying that the brain obsessed don't even realise that they are simply interpreting a scientific finding (the body image) in one of many ways, their own particular interpretation is taken to be a fact: "we now know that is false". No we don't! At least there are humans who can see this; as Max Velmans (2007, p.356) puts it:

Has science discovered that (despite appearances) pains are really in the brain as Searle suggests? It is true of course that science has discovered *representations* of the body in the brain, for example, a tactile mapping of the body surface distributed over the somatosensory cortex (SSC). However, no scientist has observed actual body sensations to be in the brain, and no scientist ever will

I will refer to the view that *all of the 'feeling states' that a human becomes aware of are located in their brain* as 'brain-centricism', and its advocates as 'braincentrics'. If one has this view and is also a physicalist then one is a 'physicalist braincentric'. If one has the 'world view' of 'physicalist braincentricism' then one will believe that non-brain parts of the human body *must* be wholly devoid of 'feeling states' and solely be the location of states such as (non-feeling) tissue damage.[57] According to this view it is the way in which a brain 'perceives' non-feeling states of the body which results in 'feeling states'. An advocate of this view is David Armstrong (1963) who claims that:

> to say the sensation of heat is in the hand is only to say that we believe, or are inclined to believe, that the bodily cause of the sensation is in the hand...something necessarily distinct from the sensation, its putative cause, is in the hand.

Another advocate is Michael Tye (1997, p.333) who claims that: "Pains...are sensory representations of tissue damage." According to this view there is a process of association whereby one matches 'feeling states' with bodily causes and thereby fallaciously comes to believe that the feeling itself is located where its associated cause is located. For example, if one becomes aware of 'feeling states' of pain at the same time as one sees a car driving over one's foot then one *assumes* that the pain is in one's foot, but it is *actually* in one's brain.

It is important to realise that the 'physicalist braincentric' position entails two separate claims. Firstly, that brains are the *only* parts of the universe that contain 'feeling states'. Secondly, that when it appears to one that there is a 'feeling state' in one's foot that it is actually in one's brain. Clearly, it is possible to

---

[57] Of course, in believing such a thing one will have created the seemingly intractable 'problem of consciousness' (the *problem of 'what-it-is-likeness'=awareness*) which we encountered in *Chapter Four*.

accept the second claim and reject the first; indeed, this is a widespread position within the panawareness/panpsychist 'world view'. According to this view one's body is pervaded with 'feeling states' but *one only ever becomes aware of* the 'feeling states' which are located in one's brain. It should be remembered that our consideration of where 'feeling states' are located is solely concerned with the 'feeling states' *that one becomes aware of.*

Let us recap. The 'physicalist braincentric' believes that all of the 'feeling states' that one becomes aware of are located in one's brain *and* that there are no 'feeling states' in any non-brain part of the universe. However, 'braincentricism' is a broader position of which 'physicalist braincentricism' is one variety. The belief of the 'physicalist braincentric' that all of the 'feeling states' that a human becomes aware of are located in the brain of that human is a belief that is shared with other 'world views'. One can believe that the entire universe is pervaded by 'feeling states' but that a human only ever becomes aware of the 'feeling states' which are located in their brain (a human only has access to these 'feeling states'). It is this latter belief – that a human only ever becomes aware of the 'feeling states' which are located in their brain – that is the mark of a 'braincentric'.

5.3.2 *The non-location of 'feeling states'*

The second possibility is that the 'feeling states' that a human becomes aware of do not have a location because they are attributes of a non-spatial mind. We saw that the 'braincentric' takes the claim *I have an unpleasant feeling in my finger* and reinterprets it as *there is a disturbance in my finger which causes an unpleasant feeling to be located in my brain.* In contrast, according to the non-location view, 'feeling states' are simply modes of thinking of a mind which has no location. This is the view of Descartes (1996, p.60) who claimed that:

> When the nerves are pulled in the foot, they in turn pull on inner parts of the brain to which they are attached, and produce a certain motion in them; and nature has laid it down that this motion should produce in the mind a sensation of pain, as occurring in the foot.

5.3.3 The location of 'feeling states' throughout the brain and non-brain body

The third possibility is that the 'feeling states' that a human becomes aware of are sometimes located in their brain and they are sometimes located in their non-brain body. This is clearly the view that most accords with common sense. It sometimes seems blindingly obvious to one that one has a 'feeling state' in the vicinity of one's brain; and it sometimes seems blindingly obvious to one that one has a 'feeling state' in the vicinity of the farthest part of one's body from one's brain (one's toes). This common sense view has recently been defended by John Hyman (2003, pp.23-4) who is bewildered as to why "the strange and grotesque idea that bodies are insensitive continues to predominate". Hyman (2003, p.5) asserts that:

> itches, aches, pains, tickles and so on – sensations of all sorts – are generally in the places we say they are. So, for example, if I say that I have an itch in the big toe on my left foot, then, by and large, that is the very place where the itch is. I think it would seem incredible to most people that philosophers deny this.

Another advocate of common sense is Truls Wyller (2005) who asserts that: "Hyman is right that the pain of my thumb is in my thumb". Another advocate of common sense is Robert Coburn (1966) who claims that: "in uttering...'I have a pain in my leg' what we are typically doing is saying where a certain pain is located in our bodies, giving the place of our pain, locating a pain, etc – PERIOD!"

Of course, to assert that the non-brain body is *not* wholly devoid of 'feeling states' is *not* to say that 'feeling states' are always located in the body exactly where one assumes them to be located; it is obviously quite compatible with this view for one to believe that one has a tingling sensation located in one's foot and for it to actually be located in one's knee. Indeed, it is well known that humans occasionally make mistakes, so *it is to be expected* that such errors will sometimes occur.

5.3.4 The location of 'feeling states' outside of the brain and body

The fourth possibility is that the 'feeling states' that a human becomes aware of are located outside of the brain and body of that human. The 'braincentric' will, no doubt, find this possibility to be highly implausible; for, the 'braincentric' assumes that the brain is an impermeable barrier which cannot be breached – one is stuck inside one's brain and thus cannot become aware of 'feeling states' which are located outside of one's brain.

However, if one rejects 'braincentricism' and believes that one can become aware of 'feeling states' which are located in one's foot, then there is seemingly no reason to deny the possibility that one could become aware of 'feeling states' which are located in the foot of a distantly located human (or, if one is a panwhat-it-is-likeness advocate, located in any part of one's surroundings). Should one unquestioningly assume that one's body is an impermeable barrier – that one can become aware of 'feeling states' which are located in one's foot, but that it is impossible for one to become aware of 'feeling states' which are located outside of one's body? Of course not, there is no evidence that such an impermeable barrier exists; indeed, as we will see in *Section 5.3.6,* there are a multitude of 'world views' which entail that the body is not an impermeable barrier. There is also increasing scientific evidence that the entire universe is very deeply interconnected.[58]

This position is alluded to in Charles Dickens novel *Hard Times* when Mrs Gradgrind is asked if she is in pain whilst on her sickbed. She replies (Dickens, 1969, p.224):

> *I think there is a pain somewhere in the room, but I couldn't positively say that I have got it.*

What could this assertion possibly mean? There is clearly a sense in which Mrs Gradgrind did have the pain because she was aware of the existence of the pain; in this sense to have a

---

[58] If one finds it implausible that a human can become aware of feeling states which are located outside of their body then this possibility should appear more plausible after we have explored the links between awareness and 'feelings states' in *Chapter Six.* The scientific evidence for interconnectedness primarily comes from quantum entanglement. Furthermore, it is worth recalling that there is some evidence for telepathy; scientific studies, particularly with twins, have concluded that humans can have 'telepathic' access to the states located within other humans (i.e. one's body *is not* an impermeable barrier).

'feeling state' *is* simply to be aware of its existence. However, there is another issue – the issue of the location of the pain. It is possible that the pain was not located in Mrs Gradgrind – not in her brain or non-brain body – this is what she means when she asserts that "I couldn't positively say that I have got it". The pain could have been located in the body of the human asking her the question.

It should be stressed that this possibility concerning the location of 'feeling states' is obviously not a complete account. It is hard to imagine anyone defending the view that *all* of the 'feeling states' that a human becomes aware of are located outside of that human. When one becomes aware of a painful 'feeling state' in one's foot at the same time as one sees a car driving over one's foot, it would be very odd to believe that the painful 'feeling state' was located in another human. However, in a particular instance – as with Mrs Gradgrind – when there is no obvious cause of a 'feeling state' that one becomes aware of, then the possibility that it is located outside of one should be taken seriously.

5.3.5 Why is the issue of the location of 'feeling states' important?

We have considered various views concerning where the 'feeling states' that a human becomes aware of are located. In the next section we will be comparing these theories in order to see which of them is the most plausible. The purpose of this section is to briefly remind ourselves, and further explore, why this issue is an important one.

Recall that our underlying objective is to explore the relationship between humans and their surroundings from an evolutionary perspective. One of the main questions this entails is: Are 'feeling states' states which evolve or do they pervade the universe? If one believes that the 'feeling states' that a human becomes aware of are located in a non-spatial mind, and that humans are the only parts of the universe to have evolved such non-spatial minds, then clearly one will believe that there is a great chasm between humans and the surroundings which evolved them. Similarly, if one is a 'physicalist braincentric' then one will believe that there is a great chasm between humans and the surroundings which evolved them (strictly the 'physicalist braincentric' believes that there is a great chasm between *human brains* and the surroundings which evolved these brains).[59] The existence of such chasms is not intellectually appealing.

The common sense position is that the 'feeling states' that a human becomes aware of are often located in the non-brain body of that human. This position opens the door to a view of the universe in which a chasm does not exist. One can believe that the brain of a human generates awareness and that 'feelings states' exist throughout the body of a human. According to this view a human becomes aware of a distantly located 'feeling state' when the awareness which is located in the brain of this human becomes aware of the 'feeling state' which is located in their foot.[60] If one has this view then it is easy to see how one can believe that a human can become aware of a *slightly more*

---

[59] Even more strictly, the 'physicalist braincentric' might believe that the 'great chasm' is between human brains/some non-human animal brains and the surroundings which evolved these brains.

[60] Various views concerning the links between awareness and 'feeling states' are considered in the next chapter.

distantly located 'feeling state' which is located outside of their body (indeed many 'feeling states' outside of a human's body would be *less* distantly located from their awareness than those in their foot; for example, when they are kissing another human). The only thing that one needs to believe in order to make such an extension is that the body of a human is not an impermeable barrier.

Recall that according to the 'panwhat-it-is-likeness' advocate the universe is pervaded with 'what-it-is-likeness' and it is only in certain parts of the universe that awareness evolves. Given our present discussion there are two paths that the panwhat-it-is-likeness advocate could take. Firstly, they could take the path outlined in the previous paragraph according to which the 'feeling states' that a human becomes aware of are located throughout their body. Secondly, they could believe that the entire universe is pervaded with 'feeling states', but that a human can only ever become aware of the 'feeling states' which are located in their brain.

The 'panwhat-it-is-likeness' advocate could be tempted to take the second path. For, whilst believing that 'feeling states' cannot evolve out of arrangements of the universe that are wholly devoid of feeling, it is seemingly also plausible that human brains (and some non-human animal brains) have a *high level* of 'feeling states'.[61] It is perhaps natural to believe that the 'feeling states' that a human becomes aware of must be much more complex/superior/intense than the 'feeling states' that exist in parts of the universe such as stones and atoms, and

---

[61] We have already seen that the two claims in this sentence are held by many panawareness advocates. See the Strawson quotes in *Section 3.5.2,* and the James, Laszlo, Bohm, Strawson and Griffin quotes in *Section 3.5.4.*

that these *high level* human 'feeling states' are products of the human brain. According to this view stones, raindrops and human legs contain *low level* 'feeling states'.

However, there is a problem with this taking this path. If one takes it then one faces the 'combination problem' – explaining how the *low level* 'feeling states' give rise to the *high level* 'feeling states'. There is another troubling aspect to this path. If it is held that *all* of the 'feeling states' *that one becomes aware of* are wholly located in one's brain, then, one has a good reason to doubt that the non-brain universe actually contains 'feeling states'. One's prime reason for believing in the existence of non-brain 'feeling states' will be that one does not believe that 'feeling states' can emerge out of states which are wholly devoid of feeling (and as we saw in *Section 5.3.4* this belief is ultimately a matter of faith).

The merit of the first path is that it does not create a division in the universe between *high level* 'feeling states' and *low level* 'feeling states'. According to this view the entire universe contains *same level* 'feeling states' and only some arrangements of the universe have awareness of these 'feeling states'. To assert that the entire universe has *same level* 'feeling states' is obviously *not* to assert that the entire universe has the same level of feeling *at a particular moment*; it is just to assert that there is no hierarchy of 'feeling states' with brains near the top and atoms near the bottom. According to this view there is no 'combination problem' to be solved. This is because *low level* feelings in parts of the universe such as atoms, and the foot of a human, do not combine to form *high level* feelings in the human brain; there are simply *same level* 'feeling states' throughout the universe. There is also no combination of awareness because there is no such thing as *low level* awareness and *high level* awareness; it is simply the case that there is level-less awareness in some parts of the universe, parts such as human brains.[62]

So, the issue of where the 'feeling states' that a human becomes aware of are located is of great importance. If such states are solely located in the brain of that human (braincentricism) then there is either a great chasm between humans and their surroundings (physicalist braincentricism), or there is an extremely difficult problem which needs to be solved (the combination problem). However, if the 'feeling states' that a human becomes aware of pervade the body of that human then this opens the door to a panwhat-it-is-likeness view in which there is no great chasm and no combination problem.

5.3.6 Comparing the different views of 'feeling state' location

Four different views of 'feeling state' location have been outlined – the 'feeling states' that a human becomes aware of exist solely in their brain, in both their brain and body, outside of their brain and body, or they don't exist at any location. As has already been stated, the view that a human's 'feeling states' are located outside of their brain and body does not purport to be a complete account of 'feeling state' location. The possibility that *some* of the 'feeling states' that a human becomes aware of are located outside of their body should be acknowledged to be a possibility. This 'peripheral' view of 'feeling state' location is clearly compatible with both the 'solely-brain location' and the 'brain and non-brain body location' views.

Let us consider the view that 'feeling states' exist but that they don't have a location. At first glance this view seems to be

---

[62] I make the case for 'level-less' awareness in *Section 6.8.*

very implausible. To say that a tree exists but that it doesn't exist at a location would be incoherent. To say that the Loch Ness Monster doesn't exist but that it is located in Loch Ness is just as incoherent. Why should it be any different when it comes to 'feeling states'? According to common sense if something exists it exists somewhere (at some location). If something doesn't exist then it doesn't exist anywhere (it has no location). From an evolutionary perspective this view is also unappealing. 'Physicalist braincentricism' creates an unappealing chasm between humans and the surroundings which evolved them; the non-spatial mind view creates an even bigger chasm than 'physicalist braincentricism' by adding another layer of separation between humans and their surroundings. According to this view humans have states that are non-spatial whereas the surroundings which evolved them are wholly spatial. This hypothesised emergence of the non-spatial from the spatial seems just as likely as the emergence of the extended out of the wholly unextended – and recall that Strawson claims that such a possibility "should be rejected as absurd" (*Section 3.5.2*).

So, in the name of plausibility, let us assume that there are two possibilities for our *central* theory of the location of the 'feeling states' that a human becomes aware of. A human's 'feeling states' are solely located in their brain, or they are located in both their brain and their non-brain body. How are we to judge between these two theories? If the basis of judgement is what seems to be obviously true on the basis of one's lived life, then one would surely assert that 'feeling states' are located in both one's brain and one's non-brain body. It is a frequent occurrence to become aware of 'aching' in one's arms, 'tingling' in one's legs, 'tastes' in one's mouth, and a plethora of other 'feeling states' located *throughout one's body*. Of course,

201

as we have seen, the 'braincentric' argues that humans are deluded. Why is this? One reason is the 'body image' which we have already discussed. Another reason is the 'phantom limb' phenomenon. This is clear from the following assertions from E. J. Lowe and H. Hudson:

> We speak as if pains were located 'in' various parts of our bodies, such as our backs or our toes. And yet it is difficult to interpret this manner of speaking literally, especially when we consider the so-called 'phantom limb' phenomena experienced by some amputees. Sometimes, a person who has had a leg amputated continues feeling pains and other sensations 'in' the amputated leg, but it is difficult to suppose that what such a person is aware of is anything literally *located* where his leg would have been, since that part of space may apparently contain nothing but thin air.
>
> (Lowe, 2000, p.116)

> If my arm or my finger is amputated I may still feel the same ache or throbbing as I felt before it was ampu- tated... This shows, I think, that the aches, throbs, or twinges one gets in various parts of one's person cannot be properly regarded as something that happens to the parts in question.
>
> (Hudson, 1961)

So, some humans seem to be convinced that the 'phantom limb' phenomenon means that all of the 'feeling states' that a human becomes aware of are solely located in that human's brain. However, there really is no good reason why one should accept that this is so. Let us break down the argument into four stages:

> P1 One believes that one has a 'feeling state' located *in* one's right arm

> P2 One has *no* right arm

> C1 The 'feeling state' is in one's brain

> C2 All 'feeling states' that all humans become aware of are solely located in their brains

The two premises are fine but they justify neither of the conclusions. There is no compelling reason why one should accept C1 – one could equally hold that the 'feeling state' is located in another part of one's non-brain body (such as the nerve endings of the amputated limb), or even that it is located *outside* of one. In other words, the acceptance that one can be wrong about where one's 'feeling states' are located doesn't automatically lead to the conclusion that one's non-brain body is wholly devoid of 'feeling states'. It is perfectly reasonable to hold that *sometimes* 'feeling states' are located in one's body exactly where one believes them to be and that *sometimes* they are not so located.

Furthermore, even if one accepted C1 then there is no good reason why one should accept C2. The loss of one's right arm could cause a rewiring of neural circuitry which causes 'feeling states' in one's brain which *appear* to one to be where one's right arm used to be; but this clearly doesn't mean that the 'feeling states' which appear to one to be in one's left leg are also located in one's brain. As Hyman (2003, p.18) puts it:

> The phantom limb case shows that if it feels to me as if my leg is hurting, it does not follow that my leg is hurting, since my leg may have been amputated, and no longer exist. But this does not imply that when my leg *is* hurting, the pain is not located in my leg.

The defender of the view that all of one's 'feeling states' are located in one's brain might resort to neuroscience and assert that when one believes that one has a 'feeling state' located in one's right arm that the 'feeling state' can be 'seen' in the brain. This would be an unacceptable assertion; for, it is perfectly intelligible to assert that the brain contains states of *awareness of* 'feeling states', whilst the 'feeling states' *themselves* are located in one's non-brain body. If this is so, then there would be neural correlates of 'feeling states', but the 'feeling states' themselves could be located in the non-brain body. According to this view general anaesthetic does not take away 'feeling states'; rather, general anaesthetic takes away the brain-located awareness of 'feeling states', whilst the 'feeling states' themselves are still located in the body.

We have, so far, assumed that Lowe is correct when he asserts that: "it is difficult to suppose that what such a person is aware of is anything literally *located* where his leg would have been, since that part of space may apparently contain nothing but thin air." There are those who disagree and believe that this is *not* something which is difficult to suppose; indeed, their 'world view' entails that if one has a limb removed then one *should expect* to become aware of 'feeling states' which are located in the area where one's limb used to be. There are numerous 'world views' which entail that a human can become aware of 'feeling states' which are located where a limb used to be; the only requirement is that the 'world view' entails that humans have an aura. This belief can be found in Neo-Platonism, Archeosophy, Kabala, Jainism, Buddhism, Hinduism, Theosophy, Anthroposophy, and many other 'world views'.

If one believes that humans have an aura then one believes that when a human has a limb amputated that the aura remains where the limb used to be; contra-Lowe it is not "thin air". Many humans claim to be able to perceive the auras which surround all living things.[63] Given the constraints on the human perceptual apparatus which we have considered, and given that the 'surroundings in-themselves' are an objectless 'blobject', we should not be surprised if an aura exists and that where many humans see 'thin air', some humans perceive an aura. The aura is hypothesised to be a more fundamental part of a human than their limbs (the aura is the part of the universe which the limbs grow into), but it is the limbs which are the part of a human which is easiest to perceive by other humans. If one believes in an aura then one will probably believe that 'feeling states' are

---

[63] It is interesting that auras have only been observed surrounding the 'biological' part of the universe. This ties in nicely with my claim, in *Section 2.5,* that the 'biological' part of the universe is a state of 'excitation' or 'exhilaration'. The aura could be an observable sign of such 'excitation'.

located in this 'spiritual' part of a human rather than (or in addition to) their 'easily observable body'. So, it would be expected that when a human loses a limb that they would still become aware of the 'feelings states' which are located in the aura which exists in the area where the limb used to be.[64]

Let us conclude. There would need to be a very compelling reason to make one accept that one is deluded – that one's non-brain body is actually wholly devoid of 'feeling states' and that all of one's 'feeling states' are actually solely located in one's brain. Our conclusion is that no such compelling reason exists; indeed, we have seen that there are very good reasons to believe that it is the 'braincentric' who is deluded. Let us let Wittgenstein (1980, p.12) have the final word on the matter:

> One knows the position of one's limbs... Just as one also knows the place of a sensation (pain) in the body.

5.4 The 'what-it-is-likeness' of the visual sense

Recall that there are two groups of 'what-it-is-likeness' – 'feeling states' and the 'sensory what-it-is-likeness' of the two senses. In the previous section we considered the location of the type of 'what-it-is-likeness' that is 'feeling states'. In this section our objective is to consider the location and pervasiveness of the

---

[64] There is a slightly more fruity possibility according to which the aura where the limb used to be is still correlated to the stuff which constituted the limb (quantum entanglement entails that this would be the case). If this is so, then the feeling states in this stuff could be 'transferred' to the aura where the limb used to be, rather than originating in the aura itself.

'what-it-is-likeness' of the first of the two senses – visual 'what-it-is-likeness'. Let us keep in mind that the reason for this consideration is that we are continuing to explore the relationship between humans and their surroundings from an evolutionary perspective, and our current concern is to explore this relationship from the perspective of the phenomena of 'what-it-is-likeness' and the human senses. Does the 'what-it-is-likeness' of the visual sense evolve in humans? If so, then there are grounds for concluding that humans are differentiated from the surroundings which evolved them through the phenomenon of 'what-it-is-likeness'. Alternatively, did the 'what-it-is-likeness' of the visual sense exist before humans evolved? Does it pervade the universe? If so, then visual 'what-it-is-likeness' clearly does not differentiate humans from their surroundings.

As we noted in *Section 5.1* the 'what-it-is-likeness' of the visual sense is the phenomenon of colour.[65] I will take it for granted that you understand what I mean by the word 'colour'. Yesterday I went into the garden and I perceived a tomato; this tomato had a particular shape and it had a particular colour – a particular shade of 'redness'; this 'redness' = colour. Unfortunately, some humans purport to be talking about colour when they are actually only talking about 'light waves which are themselves *not coloured*', or 'properties of objects which are themselves *not coloured*'. I will simply be using the word colour to refer to the phenomenon of colour. The *central question*

---

[65] Some people might just assume that the 'what-it-is-likeness' of colour requires a visual sensing or a perceiver of colour. But the 'what-it-is-likeness' of colour is colour, so if colour exists in the unperceived world then the 'what-it-is-likeness' of colour exists in the absence of a visual sensing or a perceiver of colour.

which is to be addressed in this section is: Does the unperceived world contain colour?

Let us consider the following scenario. At a particular time – t_1 – I visually perceive a cup that appears to be coloured and located about one metre in front of me; if describing the colour I would call it green. A moment later at t_2 I close my eyes and am unable to see the cup. A moment later at t_3 I open my eyes and I again see a cup that appears to be exactly the same colour as before; let us assume that this is so. This scenario raises the following question: *At t_2 was the cup coloured or was it wholly devoid of colour?*

This question requires a little refinement, for at t_1 and t_3 whilst *the cup* appears to me to be green this could be due to the properties of *the gap* between me and the cup. Our central question is whether the unperceived world contains colour so we need to consider whether the unperceived world contains colours in objects and/or in the gaps between objects. The term 'cup-gap' will be used to refer to the area of my surroundings that includes both the cup and the gap between me and the cup. It might help to visualise this; the 'cup-gap' starts at the outer edge of my eyes where it is narrow and extends outwards until the cup is reached; it is a cone-like shape which only encompasses the area between my eyes and the cup, and the cup itself. In order to address our central question the following question will be addressed: *At t_2 was the cup-gap coloured or was it wholly devoid of colour?* There are four possible answers to this question:

1   At t_2 the cup-gap was wholly devoid of colour.

2   At t_2 the cup-gap was coloured the same colour as the colour which I perceived at t_1 and t_3 (green).

3   At t_2 the cup-gap was coloured a different colour to the colour which I perceived at t_1 and t_3.

4   At t_2 the cup-gap was coloured green and lots of other colours.

These four possibilities will be considered in the next four sections and will be followed by a summary in *Section 5.4.5;* then, in *Section 5.4.6,* I draw some conclusions. It is worth keeping in mind that my objective in the next four sections is to outline each of the four positions above; my objective is not to provide a detailed exegesis of the views of a particular philosopher. I mention this because there is often much debate as to which views some philosophers actually held, and I don't want to get bogged down into a debate as to whether X believed Y or Z (or both Y and Z at different times or in different places in their work); so if one believes that X has a different view to that which is ascribed below that is fine. My interest is simply in the four different positions that it is possible for *any human* to adopt.

5.4.1 At t_2 the cup-gap was wholly devoid of colour

Let us refer to the view that at t_2 the cup-gap was wholly devoid of colour as 'dispositionalism'. According to this view, at t_2 the cup-gap simply contains the power or disposition to appear green when perceived. One of the first advocates of this view is Galileo (cited in Thompson, 1995, p.19) who asserted:

> I think that tastes, odors, colors and so on are no more
> than mere names so far as the object in which we place
> them is concerned, and that they reside only in the con-
> sciousness. Hence if the living creatures were removed,
> all these qualities would be wiped away and annihilated.

Other notable advocates of this view are typically taken to
be Robert Boyle, Descartes and John Locke; it is Locke who
provides the fullest elucidation of the view. He argues that the
unperceived world contains *Primary Qualities* and that the
particular arrangements of these *Primary Qualities* give rise to
certain powers – *Secondary Qualities*. These powers are located
in objects and they cause particular colours to be seen when
perceived. So, when a perceiver perceives a particular arrangement of wholly uncoloured *Primary Qualities* then they will
perceive a particular colour. In Locke's words:

> Qualities thus considered in Bodies are, First such as are
> utterly inseparable from the Body, in what estate soever
> it be; such as in all the alterations and changes it suffers,
> all the force can be used upon it, it constantly keeps; and
> such as Sense constantly finds in every particle of Matter,
> which has the bulk enough to be perceived, and the Mind
> finds inseparable from every particle of Matter, though
> less than to make itself singly be perceived by our
> Senses... These I call *original* or *primary* Qualities of
> Body, which I think we may observe to produce simple
> ideas in us, *viz.* Solidity, Extension, Figure, Motion, or
> Rest, and Number.

> *2dly*, Such *Qualities*, which in truth are nothing in the
> Objects themselves, but Powers to produce various

Sensations in us by their *primary qualities, i.e.* by the Bulk, Figure, Texture, and Motion of their insensible parts, as Colours, Sounds, Tastes, *etc.* These I call *secondary Qualities.*

(Locke, 1975, %9-10)

The *Ideas of primary Qualities* of Bodies *are Resemblances* of them, and their Patterns do really exist in the Bodies themselves; but the *Ideas, produced* in us *by* these *Secondary Qualities, have no resemblance* of them at all.

(Locke, 1975, %15)

Another notable advocate of this position is Isaac Newton who established that light consists of rays that differ in refrangibility – the degree to which they are bent. He theorised that the colour of the cup that I perceive results from the way the cup differentially reflects these light rays. Newton's experiments led him to claim that (1952, pp.124-5):

The homogeneal Light and Rays which appear red, or rather make Objects appear so, I call Rubrifick or Red-making...the Rays to speak properly are not coloured. In them there is nothing else than a certain Power and Disposition to stir up a Sensation of this or that Colour.

The dispositionalist could believe that a complete analysis of the cup-gap would reveal that I am perceiving green; that is to say, the 'green-making' properties are perceiver-independent. The second option for the dispositionalist is to deny that this is

211

so. C. L. Hardin urges dispositionalists to adopt this latter option.

Hardin (1988, p.4) argues that there are a plethora of physical properties of objects which influence their perceived colour and that therefore: "it would be in vain to suppose that objects sharing a common color resemble one another in physical structure." Furthermore, Hardin rejects the idea that there could be a family of diverse physical structures which share a disposition to radiate light of a particular character from their surfaces. This is because: "it is unlikely that any two things chosen at random which look to have the same blue color under normal conditions will have identical reflection spectra, let alone identical spectra of the other sorts" (Hardin, 1988, p.7). These considerations cause Hardin (1988, p.112) to conclude that colours cannot be located in the cup-gap: "We are to be eliminativists with respect to colour as a property of objects"; they also cause Hardin to conclude that the particular colours which I perceive are determined by properties within me (1988, p.xxi): "At the present moment there isn't the slightest reason to think that there is a set of external physical properties that is the analog of the fourfold structure of the colors that we experience". According to this view a *complete* analysis of the properties of the cup-gap at t_1 would *not* reveal which colour I perceive.

Dispositionalism is often asserted to be the *only* intellectually respectable position that a contemporary human can hold. For example, Simon Baron-Cohen (1996, p. xi) asserts:

> Just as common sense is the faculty that tells us that the world is flat, so too it tells us many other things that are equally unreliable. It tells us, for example, that color is out

there in the world, an independent property of the objects that we live among. But scientific investigations have led us, logical step by logical step, to escape our fanatically insistent, inelastic intuitions. As a result, we know that color is not already out there, an inherent attribute of objects.

5.4.2 At t₂ the cup-gap was coloured the same colour as the colour which I perceived at t₁ and t₃ (green)

Our second possibility is that at t_2 the cup-gap was coloured and it was coloured *solely* the colour that it appears to me to be at t_1 and t_3. So, according to this view the cup-gap was coloured green at t_1, t_2 and t_3, it is just that at t_2 I was unable to perceive that the cup-gap was green. Let us refer to this view as the 'Intrinsic Property' view. This view of colour was prevalent two thousand years ago; according to Philip Ball (2008, p.39):

> the ancient Greeks...regard[ed] colour as an intrinsic property, requiring light only to activate it like electricity activating a light-bulb.

One could believe either that light 'activates' colours that are in objects, or that objects are always coloured (and don't need 'activating'). On this latter view colours exist but the absence of light means that humans are unable to perceive them (it is *humans* that need 'activating'). The belief that the colour green can be identified with properties of the cup-gap is held by the 'primitivists' who assert that colours are *sui generis*. For example, John Campbell (2001, p.178) argues that colours supervene on physical properties:

> On this view, redness, for example, is not a disposition to produce experiences in us. It is, rather, the ground of such a disposition. But that is not because redness is a microphysical property – the real nature of the property is, rather, transparent to us... colours are supervenient upon physical properties, if only in the minimal sense that two possible worlds which share all their physical characteristics cannot be differently coloured.

Campbell stresses that an 'objective' description of the world can be one in which 'mind-independent properties' – such as colours – are absent; he sees the failure to understand this fact as the central objection which is raised against the primitivist. On this account, as colours supervene on physical properties, the cup-gap at t_2 will be solely coloured green; this means that there is no possibility of colour spectrum inversion:

> the qualitative character of a colour-experience is inherited from the qualitative character of the colour. It depends on which colour-tracking capacity is being exercised in having the experience. So if you and I are tracking the same colours, our colour-experiences are qualitatively identical. This view does not allow for the hypothesis of spectrum inversion
>
> (Campbell, 2001, p.189)

5.4.3 At t₂ the cup-gap was coloured a different colour to the colour which I perceived at t₁ and t₃

Let us now consider the odd-sounding possibility that at t_2 the cup-gap was coloured but that it wasn't coloured the colour which I perceived at t_1 and t_3 (green). At t_2 the cup-gap could have been coloured a colour such as red, it could have been coloured a colour which humans are unable to perceive and therefore have no name for, or it could have been multiply non-green coloured.

According to this view not only was the cup-gap at t_2 a completely different colour to the colour which I perceived at t_1 and t_3, but it is also the case that at t_1 and t_3 the cup-gap was a completely different colour to the colour that I perceived at t_1 and t_3. In other words, the universe in-itself is always determinately coloured, but the colours which perceivers perceive are different to these colours.

An account of this kind has seemingly been recently defended by Peter Unger. Unger's motivation for his account is his disagreement with Locke (Unger, 2006, p.168):

> Locke didn't allow that basic physical reality should have any Qualitative Color... we should liberate ourselves from the Denial, as stultifying as it's unnecessary.

From this starting point Unger claims that physical reality could have definite colours but perceivers would be unable to perceive what these colours are. So:

> In this actual world, and in this present Eon of the world, it seems quite certain, at least to me, that there aren't any sentient beings able to perceive how it is Qualitatively

> with a Spatially Extensibly Qualitied physical object...
> For example, even if it should be that, in our World and
> in our Eon, every electron is the very same Absolutely
> Specific Shade of Extensible Color, say Transparent Blue,
> there won't be any perceivers, in our World, and in our
> Eon, who're ever able to perceive any of the electrons to
> be (Transparent) Blue.
>
> (Unger, 2006, p.112)

It is clearly possible that there are two very different types
of colours, those which are located in the universe in-itself and
those which a perceiver perceives when they perceive the
universe. Unger claims that the 'basic concreta' of the universe
could be coloured a colour that a human cannot perceive. This
means that at t_2 the cup-gap was coloured a colour 'we-know-
not-what', but it is not coloured green:

> Supposing that our hypothesis is right...it's likely that
> actual physical things *won't* be Spatially Extensibly
> Colored in any of the Absolute Specific Ways that, quite
> directly and clearly, *we can conceive them to be Col-
> ored*... But, still, the basic concreta, in our actual world,
> will be Extensibly Spatially Colored concreta. Their space
> will be pervaded by less available Absolutely Specific
> Color Qualities... Let's call one of them Col, another Lor.
> So, though it may be that no real spatial concreta are
> Qualitied Redly, many may be Qualitied Lorly.
>
> (Unger, 2006, p.168)

According to the view outlined in this section – a view
which I will call the 'we-know-not-what' view – at t_2 (and t_1 and

t₃) the cup-gap is coloured a particular colour (or colours) we-know-not-what, but it is not coloured green.

5.4.4 At t₂ the cup-gap was coloured green and lots of other colours

Finally, let us consider the possibility that at t₂ the cup-gap contained a multitude of colours including the colour which I perceived at t₁ and t₃ (green). In other words, the cup-gap (at t₁, t₂ and t₃) was a plethora of intermingling colours. This view is compatible with the idea that the universe in-itself is not determinately coloured – rather, every point in the universe contains a plethora of colours; so, whilst a perceiver can only perceive one colour in a particular location at any given time, the universe actually contains a multitude of colours at this location. This view is also compatible with the idea that every point in the universe is determinately coloured.

There are two different accounts which seek to explain why I perceive green in the cup-gap when the cup-gap actually contains green and a plethora of other colours. According to the first account it is properties in the cup-gap which determine that I perceive green. So, Aristotle (1980, pp.17-9) argues that:

> for when light falls on something, and, being tinted by it, becomes reddish or greenish, and then the reflected light falls on another colour, being again mixed by it, it takes on still another mixture of colour. And being affected in this way, continually but imperceptibly, it sometimes reaches the eyes as a mixture of many colours, but producing the sensation of the most predominant

According to the second account it is the structure of my visual perceptual apparatus which causes me to perceive the cup-gap as green rather than as the intricate interfusion of a plethora of colours that the cup-gap actually is. In other words, the green that I perceive is located in the cup-gap, whilst my visual perceptual apparatus operates as a filtering device which has no access to the non-green colours in the cup-gap. So, it is the structure of my visual perceptual apparatus that determines which colours I perceive. This means that if instead of me perceiving the cup-gap at t_1 a different perceiver took my place that perceiver could perceive a non-green colour (in contrast to the first Aristotelian account). As most humans have very similar visual perceptual structures one would expect that most humans would perceive the same colour; it is more likely that an entity with a very different visual perceptual structure would perceive a non-green colour.

This account of colour is concordant with the views of physicist Frank Wilczek, who philosopher Evan Thompson (1995, p.16) claims is "a leading physicist". According to Wilczek and Devine (1988, p.6, p.9):

> Science early discovered that light "in itself" has a much richer structure than our sense of vision reveals. It therefore becomes necessary to distinguish <u>physical colour</u> from <u>sensory colour</u>. It is possible to predict the sensation people will report when a given bundle of light rays impinges on their eyes – that is, the sensory colour – in terms of the physical colours present in the rays. On the other hand, there are combinations of light that are definitely different but look the same. Physical colour is, in this precise sense, more fundamental than sensory colour, even though the latter is what we actually see.

as far as we know there is a continuous infinity of possible pure colours. Some light is refracted through each of a continuum of slightly different angles emerging from the prism, and the light at each angle represents a different pure colour.

As the complete range of sensory colours that a human can perceive arises from different combinations of only three pure colours, whilst the universe itself contains a "continuous infinity of possible pure colours", it follows that the universe in-itself is truly an intermingling web of a multitude of colours only a few of which any one perceiver can possibly ever perceive. In Wilczek and Devine's (1988, p.10) words: "we are all colour blind. At best, we perceive three averages from an infinite manifold of physical colours."

So, whilst there are two variants of the 'Infinite Manifold' view, *all* advocates of this view assert that at t_2 (and t_1 and t_3) the cup-gap contains green and a multitude of other colours.

5.4.5 *Summary*

We have considered four possible answers to the following question:

> *At t_2 was the cup-gap coloured or was it wholly devoid of colour?*

Answer 1   Dispositionalism

The cup-gap is wholly devoid of colour, and the colour that I perceive (green) is either determined by properties of the cup-gap or by properties of my perceptual apparatus.

Answer 2   The Intrinsic Property View

The cup-gap is coloured solely green because green is an intrinsic property of the cup-gap which exists in the absence of a perceiver.

Answer 3   The 'We-Know-Not-What' View

The cup-gap is coloured a particular colour (or colours) we-know-not-what (such as 'Lorly'), and the cup-gap is wholly devoid of green.

Answer 4   The Infinite Manifold View

The cup-gap contains a multitude of colours of which green is one.

5.4.6 Conclusions

There are a number of interrelated reasons which lead to the conclusion that the *'Infinite Manifold' View* is the most plausible description of the universe. Firstly, the *Intrinsic Property View* seems to be too simplistic because there are good scientific reasons for believing that the colour that a perceiver perceives is dependent upon the structure of the perceptual apparatus of the perceiver; one wants a theory of visual perception which allows for the possibility of 'spectrum inversion'. This is why Hardin concludes that a complete analysis of the physical properties of the cup-gap could not reveal the fact that I am perceiving green. As we also saw in *Section 5.4.1*, these scientific reasons also cause Baron-Cohen to assert that the *Intrinsic Property View* must be false, and *Dispositionalism* must be correct, as: "we know that colour is not already out there".

Secondly, the *'We-Know-Not-What' View* is too extrava-gant. If the cup-gap appears to me to be green, then to *both* deny that there is green in the cup-gap *and* assert that the cup-gap is actually coloured a different colour is hard to accept. It is much more plausible to assert that if there is colour (or colours) in the cup-gap that green *is* in the cup-gap.

Thirdly, Unger seems to be onto something when he rejects *Dispositionalism* on the basis that the denial of colours to the unperceived world is "stultifying" and "unnecessary". We ordinarily take the universe to be coloured whether it is perceived or not, and would need a very good reason to deny that this is so. Furthermore, one can believe that Secondary Qualities are simply *not separable* from Primary Qualities.

Fourthly, the *Infinite Manifold View* has several advan-tages; it doesn't entail that the unperceived universe is wholly devoid of colour; it doesn't entail that the cup-gap is wholly devoid of green; and, it advocates an active role for the perceiver in determining which colour is perceived (in the second version of the view).

So, there are obvious drawbacks with the alternative views which are all overcome by the *'Infinite Manifold' View*. This view is compatible with the science of visual perception and accommodates phenomena such as spectrum inversion. In other words, Baron-Cohen makes an unacceptable jump – the perceiver-dependence of colour does not automatically lead to the conclusion that: "we know that colour is not already out there". The *'Infinite Manifold' View* also entails that the unper-ceived universe is pervaded with colour rather than being wholly devoid of colour, and that these colours include the colours that perceivers perceive. From the evolutionary perspective there are also good reasons to favour the *'Infinite Manifold' View* (at least to favour it over *'Dispositionalism'*); for, according to this view the 'what-it-is-likeness' detected by the visual sense pervades the universe and is not something which springs into existence at a certain point in the evolutionary process. Contrarily, according to *'Dispositionalism'* there is a significant difference between humans and their surroundings; for, on this view, humans contain visual 'what-it-is-likeness' whilst the overwhelming majority of their surroundings are wholly devoid of this phenomenon.

According to the *'Infinite Manifold' View,* and the *'Intrinsic Property' View,* human visual 'what-it-is-likeness' is located in the unperceived universe; it pervades the universe and existed long before there were perceivers.[66] According to the *'We-Know-Not-What' View* colours pervade the unperceived universe, but the colours that a perceiver perceives (*their* visual 'what-it-is-likeness') are not located in the unperceived universe. According to *'Dispositionalism'* the unperceived world is wholly devoid of colour/visual 'what-it-is-likeness'. If one adopts my favoured *'Infinite Manifold' View,* the *'Intrinsic Property' View,* or even the *'We-Know-Not-What' View,* then one will conclude that there is no division between humans and their surroundings in the realm of visual 'what-it-is-likeness'.

---

[66] There is an objection to the idea that human visual 'what-it-is-likeness' exists in unperceived human surroundings which is known as the 'Martian Objection'. One is asked to imagine a Martian who has visual 'what-it-is-likeness' when they connect to *soundwaves*. This thought experiment is designed to try and persuade one that visual 'what-it-is-likeness' must be created by perceivers rather than residing in their surroundings. The Martian Objection is easily rebutted by the view of humans and their surroundings that I am developing; this will become clear after I have considered the link between perception and 'what-it-is-likeness' in *Section 6.5.*

5.5 The 'what-it-is-likeness' of the auditory sense

In this section our objective is to consider the location and pervasiveness of the 'what-it-is-likeness' of the second of the two human senses – hearing. The 'what-it-is-likeness' of the auditory sense is the phenomenon of sound. There are a plethora of sounds including those that a human might refer to as chirping, whirring, coughing, singing and humming. The issue before us is where this 'what-it-is-likeness' is located, and whether it exists in the unperceived world. In other words, when one's auditory sense senses a chirping sound where is this chirping sound located? Is it located inside of one or outside of one? If the sounds that a human perceives are located outside of them (to be more precise: if the sounds are located outside of the human's perceptual apparatus) then sounds seemingly exist in the unperceived world. In order to explore this issue three possibilities will be considered. These three possibilities are similar to those we considered for the location of colour so our discussion can be brief:

1   The unperceived world is wholly devoid of sound.

2   The sounds that a perceptual apparatus perceives are located outside of that perceptual apparatus.

3   The unperceived world contains sounds but the sounds it contains are entirely different to the sounds that a perceptual apparatus perceives.

5.5.1 *The unperceived world is wholly devoid of sound*

According to this view the unperceived world contains vibrations which resonate outwards from points of origin; these vibrations are wholly devoid of sound and they are converted into sounds when they enter the auditory apparatus of a perceiver. This is the 'Dispositionalism' View which we encountered in *Section 5.4.1*. Recall that Locke, for example, claims that both colours and sounds are Secondary Qualities which do not exist in the unperceived world. Another advocate of this view is D. C. L. Maclachlan (1989, p.26) who argues that: "the sounds directly perceived are sensations of some sort produced in the observer when the sound waves strike the ear."

5.5.2 *The sounds that a perceptual apparatus perceives are located outside of that perceptual apparatus*

According to this view sound waves are not wholly devoid of sound; sound waves contain sound and the sounds that an auditory perceptual apparatus perceives exist in the absence of the perceptual apparatus. This view includes the 'Intrinsic Property' View which we encountered in the realm of colour. There are two possible candidates for the identification of sounds. Firstly, one could believe that sounds are located in the medium between the point of origin (a point of disturbance in the universe) and the auditory apparatus; this is clearly compatible with the belief that a human only ever "directly perceives" the sounds that strike their ears. Secondly, one could believe that sounds are located solely at their point of origin. Both of these views are compatible with the belief that a plethora of sounds exist in the universe which humans are unable to perceive; indeed, they seem to necessitate that this is so. Let us consider these two possibilities.

5.5.2.1 *The location of sounds in the medium between the point of origin and the perceiver*

According to this view sounds are mechanical vibrations which are transmitted by an elastic medium. The view can be traced back to Aristotle (2001, p.586) who claimed that: "sound is a particular movement of air". More recent advocates of the view are John R. Pierce and Edward E. David (1958, p.26) who assert that:

> Sound is generated by any sort of motion – by a violent explosion, by the buzzing wings of a fly, by the vibrations of cymbals struck together. Part of the energy of the motion goes into sound, which travels out in all directions, eventually striking walls, tables, chairs and, perhaps, human ears.

According to this view the sounds that a human hears can be identified with properties of the soundwaves themselves such as their frequency and amplitude. So, when my vocal cord moves this creates a disturbance in the surrounding medium and causes the particles of the medium to move in a back-and-forth motion at a given frequency and amplitude. This motion then propagates to neighbouring particles at the same frequency, whilst undergoing an energy loss that entails a decrease in amplitude. In this way sounds exist in a medium and propagate through that medium.

There are two possibilities when it comes to the relationship between sounds and a particular auditory perceptual apparatus. Firstly, one could believe that it is when propagations reach the auditory apparatus that sounds are perceived – these sounds will be determined by the exact properties of the sound-waves as they strike the ear. Secondly, one could believe that the auditory apparatus is able to perceive the sounds that exist in the propagations that are in the vicinity of the ear; in other words, the propagations do not need to reach the ear in order to be heard.

5.5.2.2 *The location of sounds at their point of origin*

According to this view sounds are processes or events that occur at the point of origin of a motion – that is, sounds exist where a vibrating object exists. So, whilst the perception of a sound requires a medium, sounds exist at their point of origin without a medium. According to this view what is perceived are *not* sound waves located in a medium, rather, distant sounds are perceived via a medium. In other words, the waves are the effects of sounds rather than themselves being sounds. An advocate of this view is Casey O'Callaghan (2009) who asserts that:

> What kind of thing is a sound? According to one view, sounds are waves...We don't, however, *experience* sounds to be just like waves...Sounds don't seem to travel through space unless the things making them do. Furthermore, the waves that arrive at your ears weave together information from different sources into a single complex pattern of vibration. However, we sometimes hear *different* sounds, which perhaps come from differ-

ent directions, at the very same time...As a result, it seems to me most plausible to hold that sounds are particular *individuals* that have or possess sensible features like pitch, timbre, and location. This account makes them in one respect more like the objects we see than like their features or attributes. If so, sounds are not properly classified with the secondary *qualities*, such as colour, taste, and smell, as Locke famously claimed

5.5.2.3 *The completeness of the identification*

Having considered two possible views concerning the identification of sounds with unperceived states of the universe, let us now consider the nature of the identification. Both of these views seek to equate a particular sound with a particular vibration. The question is whether this identification is likely to be complete. In other words, when a sound that is perceived is equated with a particular vibration could there actually be *a lot more sound* in the vibration than is perceived? This possibility is very similar to the 'Infinite Manifold' View which we considered in the realm of colour.

Recall that we are considering the possibility that the sounds which a perceptual apparatus perceives are located outside of that apparatus. Now, it is unlikely that one will believe that a perceptual apparatus perceives *all* of the sounds which are located outside of it. Contrarily, it seems that an auditory perceptual apparatus *'tunes into' a certain range of sounds.*[67] There are likely to be many more sounds in the

---

[67] This is the 'limited scope' inevitable constraint which we considered in *Section 3.1.3.*

227

immediate surroundings of an auditory perceptual apparatus than that apparatus can 'tune into'. Furthermore, it is entirely possible that there are many more sounds in a *particular vibration* than any auditory apparatus can possibly register.[68] If this is the case then any attempt to identify particular sounds with particular states of the universe would obviously be woefully incomplete; the identification would, at best, only work for one particular type of auditory perceptual apparatus. If there are more sounds in the surroundings surrounding a perceptual apparatus than that perceptual apparatus can possibly perceive then this means that one could not tell which sounds I am hearing from a complete analysis of my surroundings; one would also need information about my auditory perceptual apparatus and how it is only 'tuning into' some of the vibra-tions=sounds which exist in my surroundings. Furthermore, if there are multiple sounds in *a particular vibration* then this means that even if one had information on which vibrations my perceptual apparatus was 'tuning into' one would still not know which sounds I perceived; for, the identification of sounds with vibrations would be incomplete.

5.5.3 *The unperceived world contains sounds but the sounds it contains are entirely different to the sounds that a perceptual apparatus perceives*

This possibility resembles the 'We-Know-Not-What' View of colour. According to this view the unperceived universe is pervaded by sounds, but these sounds cannot be heard because

---

[68] In order to appreciate this possibility it might be useful to reread the discussion of Evolutionary Epistemology in *Section 3.1.3.*

an auditory apparatus transforms them in the act of perceiving. This means that there are two distinct types of sounds – sounds 'in-themselves' and perceived sounds.

Again, as with colour, this account does have some intuitive appeal when contrasted to 'Dispositionalism' – the *transformation* of sounds could be thought to be more intelligible than the *creation* of the sounded out of the wholly unsounded. However, again we have to ask: *Why should one believe this rather than the similar but less problematic view that the sounds which I perceive exist in the universe?*

5.5.4 *Conclusions concerning the location of sound*

We have been considering the location of the 'what-it-is-likeness' of the second of the two human senses: hearing. The 'what-it-is-likeness' of the auditory apparatus = sounds. Do sounds spring into existence within certain parts of the universe – such as humans – or have sounds always pervaded the universe? If auditory 'what-it-is-likeness' springs into existence within humans then the phenomenon clearly creates a division between humans and the surroundings which evolved them. Our ongoing assumption is that, from an evolutionary perspective, fewer divisions and chasms between humans and their surroundings is more intellectually appealing than more divisions and chasms. On this basis alone we have a very good reason to favour views of the location of sound which entail that there is no such division.

So, we have a good reason to prefer the idea that the sounds that a perceiver perceives are located outside of that perceiver, to the idea of 'Dispositionalism'. Even if one accepts this, we have seen that there is still much debate to be had. Is auditory 'what-it-is-likeness' located at the point of origin of a vibrating object? Is auditory 'what-it-is-likeness' located in the medium between this object and a perceiver? Is there much more auditory 'what-it-is-likeness' in human surroundings, and within particular vibrations, than humans can possibly perceive? And, could the act of perceiving sounds transform them into a new form? As with colour, I take this latter option to be a bit too fruity.

The idea that sounds exist in human surroundings, rather than springing up within humans, is a very attractive idea. It accords with common sense and unsurprisingly has a lot of philosophical supporters too. There can be both auditory 'what-it-is-likeness' in the unperceived universe and an active role for the auditory perceptual apparatus in determining which sounds are 'tuned into' (as we have seen this is compatible with 'sound inversion'). It seems that humans might not be divided from their surroundings by the phenomenon of auditory 'what-it-is-likeness'.

5.6 Conclusions

In this chapter we have been considering the relationship between humans and their surroundings from the perspective of 'what-it-is-likeness' and the human senses. I have outlined why I believe there to be two very distinct groups of 'what-it-is-likeness'. The first group is the family of 'feeling states' which includes 'bodily feelings', the 'what-it-is-likeness' component of 'emotions' and the states of 'what-it-is-likeness' that are 'touching', 'tasting' and 'smelling'. The second group is the 'sensory what-it-is-likeness' of 'seeing' and 'hearing'.

We have explored the concept of a 'sense' and the question of how many human senses there are. In accordance with the division outlined in the previous paragraph I have contended that there are only two human senses – seeing and hearing. This is because 'feeling states' arise when two parts of the universe interact at a single location, whereas 'seeing' and 'hearing' involve the sensing of distantly located surroundings. I have also contended that 'seeing' and 'hearing' are distinguished from the non-senses because they perceive distantly located 'what-it-is-likeness'.

From the evolutionary perspective we have considered the possibility that 'seeing' and 'hearing' evolved in humans; but that the family of 'feeling states' did not evolve. According to the 'panwhat-it-is-likeness' view 'feeling states' pervade the universe and always have done. According to this view there is no division between humans and their surroundings due to the existence of 'feeling states' in humans. When it comes to the 'sensory what-it-is-likeness' of the two senses I have considered various possibilities and concluded that there are good reasons to believe that both types of 'sensory what-it-is-likeness' pervade the universe and always have done. If this is right, it means that there is no division between humans and their surroundings due to the existence of 'sensory what-it-is-likeness' in humans.

So, our overall conclusion is that there are good reasons to believe that there is no division between humans and their surroundings arising from the phenomena of 'what-it-is-likeness'. All types of 'what-it-is-likeness' pervade the universe; the evolution of humans did not entail the bringing into existence of any new type of 'what-it-is-likeness'.

Chapter 6

Awareness, perception & 'what-it-is-likeness'

In *Chapter Five* we considered the question of how similar humans are to the surroundings which evolved them from the perspective of the phenomena of 'what-it-is-likeness' and the human senses. In this chapter our focus switches to the phenomena of awareness and perception. Could these phenomena create a chasm in the universe between humans and their surroundings?

In order to explore this possibility we need to consider the relationship between awareness, perception and 'what-it-is-likeness'; for, these three are often thought to be a tightly-coupled, if not inseparable, triplet. We can think of the inseparability of the triplet as a very 'thick' conceptualisation and complete separability as a very 'thin' conceptualisation. If one believes that the three phenomena are an inseparable triplet then one is more likely to believe that there is a chasm between humans and their surroundings; for, one is unlikely to believe that the triplet pervades human surroundings.

We have already explored the nature of some of these phenomena; for example, in *Section 3.2* I contended that perception does not involve concepts. I have also made the case for some 'thin' relationships between members of the triplet; so, in *Section 4.1* I contended that 'what-it-is-likeness' and 'awareness' are distinct phenomena. There is much more that needs to be said concerning both the nature of these three phenomena and the relationships between them; this is our concern in this chapter.

Our attempt to make some progress in understanding how these three phenomena are related is clearly not aided by the fact that humans are unable to perceive the interior states of their surroundings. It would clearly be intellectually unaccept-able to simply assert that what one is barred from accessing does not exist. One is barred from the interior states of a bumble bee but this does not mean that a bumble bee is not aware; one is barred from the interior states of a stone but this does not mean that a stone does not contain 'feelings states'. This 'barring' not only exists in one's surroundings it also exists within one. Does one perceive when one is not aware? Does one contain 'feelings states' when one is not aware?

In *Section 6.1* our concern is the possibility that 'what-it-is-likeness' and awareness are equivalent. Then, in *Section 6.2,* we consider the possibility that 'what-it-is-likeness' can exist without awareness. In order to complete the analysis of the relationship between 'what-it-is-likeness' and awareness, in *Section 6.3* our focus is the converse possibility that awareness can exist without 'what-it-is-likeness'. In *Section 6.4* our focus turns to the range of possible relationships that could exist between perception and awareness. This is followed, in *Section 6.5,* by a consideration of the possible relationships that could exist between perception and 'what-it-is-likeness'. A full understanding of the relationship between the three phenomena requires an appreciation of how they relate to sleep and dreaming; so, this topic is the subject of *Section 6.6*. In *Section 6.7* I draw some conclusions concerning the relationship between the three phenomena. Our focus changes slightly in

Section 6.8 where our concern is whether or not there are levels of awareness and altered states of awareness. Finally, in *Section 6.9*, I outline the 40-Hz oscillation theory of awareness.

6.1 Awareness as 'what-it-is-likeness'

A widespread contemporary view of the relationship between awareness and 'what-it-is-likeness' is that the two phenomena are the same thing; this means that 'awareness' and 'what-it-is-likeness' are words that refer to the same state of the universe. According to this view there is no such thing as a state of 'what-it-is-likeness' that is wholly devoid of awareness, and there is no such thing as a state of awareness that is wholly devoid of 'what-it-is-likeness'. In other words, a *single state* of the universe exists which is intrinsically and necessarily both 'what-it-is-likeness'-involving and awareness-involving.

Various terms are used by the advocates of this view to refer to the *'what-it-is-likeness'-involving* aspect of this singular state – 'experience', 'something it is like', 'qualia', 'feeling', 'what-it's likeness', and 'qualitative feeling'. When it comes to the *awareness-involving* aspect of this singular state it is typically referred to by the words 'awareness', 'conscious' and 'consciousness'.[69] Here are some assertions by advocates of the dominant *awareness as 'what-it-is-likeness'* paradigm:

---

[69] Following our discussion of the "problem of consciousness" in *Section 4.1* one will probably have realised that the *awareness as 'what-it-is-likeness'* view is the 'standard contemporary view of consciousness'. Advocates of this 'thick view' of consciousness are typically mystified as to the seeming intractability of the "problem of consciousness".

The most remarkable fact about the universe is that certain parts of it are conscious. Somehow nature has managed to pull the rabbit of experience out of a hat made of mere matter.

(William Seager, 1999, p.i)

We can say that a being is conscious if there is *something it is like* to be that being.

(David Chalmers, 1996, p.4))

I use the word "consciousness" to mean, roughly, experience. And I think of experience, broadly, as encompassing thinking, feeling, and the fact that a world "shows up" for us in perception.

(Alva Noe, 2009, p.8))

experience always involves some minimal awareness

(David Griffin, 1998, p.131)

Consciousness, as I understand it, is the property things have when there is *something that it is like to be them.*

(Philip Goff, 2006b, p.1))

Consciousness...is part of our oldest biological endow-
ment. Remember that we are dealing with the
phenomenon of sentience, of feeling, seeing, smelling,
and so on.

(Colin McGinn, 1999, pp.62-3)

It is a remarkable fact about consciousness...that there is
a qualitative feel to any conscious state.

(John Searle, 2005, p.202)

experience, 'consciousness', conscious experience,
'phenomenology', experiential 'what-it's likeness', feeling, sensation, explicit conscious thought as we have it
and know it at almost every waking moment. Many
words are used to denote this necessarily occurrent
(essentially non-dispositional) phenomenon, and I will
use the terms 'experience', 'experiential phenomena' and
'experientiality' to refer to it.

(Galen Strawson, 2006, p.3)

These quotes indicate the contemporary pervasiveness of
this view; it is very widely held and it encompasses humans who
otherwise espouse very diverse views. Of those quoted above we
have the 'naturalistic dualism' of Chalmers, the 'mysterianism'
of McGinn , the 'biological naturalism' of Searle, the 'substance
dualism' of Goff, and the 'panpsychism' of Strawson. Despite
their diverse views all of these humans are united by their belief

in *awareness as 'what-it-is-likeness'*. This view can be seen as entailing the following two claims:

> *Awareness is a 'what-it-is-likeness'-involving state of the universe*

> *'What-it-is-likeness' is an awareness-involving state of the universe*

It is the first of these claims that is seemingly the reason why this view has so many supporters. This is because whenever one is aware one is typically aware of 'what-it-is-likeness' – this makes it very easy to just assume that awareness is 'what-it-is-likeness'-involving; that is to say, it is easy to assume that there is a singular state of awareness='what-it-is-likeness'. However, when we turn to the second claim then there are grounds for doubting the validity of the first claim. This is because there are many reasons why one might want to reject the second claim, and if this claim is rejected then the whole *awareness as 'what-it-is-likeness'* view gets rejected. So, let us turn to the possibility of 'what-it-is-likeness' without awareness.

6.2 *'What-it-is-likeness' without awareness*

Are there good reasons to believe that 'what-it-is-likeness' can exist in the absence of awareness? There are two very different philosophical paradigms which entail the existence of 'what-it-is-likeness' without awareness. The first of these is the 'higher-order monitoring' theory of awareness, the second is 'panwhat-it-is-likeness'. There are also other reasons for believing that there can be 'what-it-is-likeness' without awareness. Let us consider each of these in turn.

6.2.1 Higher-order monitoring

In the western philosophical tradition the roots of the 'higher-order monitoring' view of awareness are usually traced back to the 'inner sense' view of John Locke. Locke (1975, p.123) asserted that: "the perception of the operations of our own mind within us...might properly enough be called internal sense." In accordance with this assertion Locke (1975, p.138) defines awareness as follows: "Consciousness is the perception of what passes in a man's own mind". Here the term 'consciousness' is simply used to refer to a state of being aware of something, of being conscious of something.[70] The term 'consciousness' is not being used to refer to 'what-it-is-likeness' as it is by all of the philosophers we encountered in the previous section.

In contemporary philosophy there are two groups of 'higher-order monitoring' theorists. Both of these groups share the belief that awareness is an inner sense that exists at a 'higher order' and is directed towards lower level states; their disagreement is whether the 'higher-order' is perception-like or thought-like. David Armstrong and William Lycan advocate Higher-Order Perception (HOP) theories, whilst David Rosenthal advocates a Higher-Order Thought (HOT) theory. Let us look briefly at these theories with our main concern being the possible existence of 'what-it-is-likeness' without awareness.

---

[70] This is the 'thin' conceptualisation of consciousness. Consciousness = awareness (and *not* consciousness = awareness = 'what-it-is-likeness').

Lycan (1996, p.14) claims that: "consciousness is the functioning of internal *attention mechanisms* directed at lower-order psychological states and events". When it comes to the link between awareness and 'what-it-is-likeness' Lycan (1996, p.43) claims that:

> the inner-sense theorist hardly need hold that monitoring does bring qualia into being. The monitoring only makes the subject aware of a quale that was there, independently, in the first place...the inner-sense theory is simply not a theory of what makes a state qualitative in the first place.

Armstrong (1980, p.15) claims that: "Consciousness is a self-scanning mechanism in the central nervous system." When it comes to the link between awareness and 'what-it-is-likeness' Armstrong (1997, pp.721-2) argues that 'what-it-is-likeness' can exist without awareness (he calls this *minimal consciousness*):

> In thinking about consciousness, it is helpful to begin at the other end and consider a totally unconscious person. Somebody in a sound, dreamless sleep may be taken as an example...A single faint sensation is not much, but if it occurs [in such a person], to that extent there is consciousness...I call consciousness in this sense "minimal" consciousness.

Armstrong claims that that perceptions of one's "environment and bodily state" (1980, p.59) can exist without awareness (he calls this *perceptual consciousness*). According to Armstrong, awareness (which he calls *introspective consciousness*) is distinct from 'what-it-is-likeness' and perception: "Without introspective consciousness, we would not be aware that we existed" (1980, p.67); "Introspective consciousness, then, is a perception-like awareness of current states and activities" (1997, p.724).

In contrast, David Rosenthal (1997, p.741) claims that the higher-order is not perception-like but that is constituted by thought: "We are conscious of something, on this model, when we have a thought about it." When it comes to the link between awareness and 'what-it-is-likeness' Rosenthal (1997, p.732) claims that:

> If we are intermittently unaware of a pain by being distracted from it, we feel the pain only intermittently; similarly with its hurting and our being in pain. Still, it is natural to speak of having had a pain that lasted throughout the day, and even to say that one was not always aware of that pain. This provides evidence that commonsense countenances the existence of nonconscious pains. Feeling pains and having them seem equivalent only because of our lack of interest in the nonconscious cases.

So, the 'higher-order monitoring' theorists believe that 'what-it-is-likeness' can exist without awareness. It is worth noting that if the monitoring were to be reconceptualised as a 'same-order' phenomenon then the arguments for the disparity between 'what-it-is-likeness' and awareness would still apply.

6.2.2 Panwhat-it-is-likeness

We have already encountered the world view of panwhat-it-is-likeness in previous chapters. According to this view the entire universe is pervaded with 'what-it-is-likeness', but only a relatively few states of the universe are states of awareness. So, the central component of the view is that 'what-it-is-likeness' can exist without awareness. Advocates of this view believe that the evolution of 'what-it-is-likeness' out of that which is wholly devoid of 'what-it-is-likeness' is deeply problematic.[71] Furthermore, they seen no reason to deny that all of human surroundings are pervaded with 'what-it-is-likeness' and thereby to create the seemingly intractable "problem of consciousness" which we encountered in *Section 4.1*.

6.2.3 Other Reasons

Antonio Damasio and Ned Block provide two different reasons for believing in the existence of 'what-it-is-likeness' without awareness. Here is what they have to say:

> There is...no evidence that we are conscious of *all* of our feelings, and much to suggest that we are not. For example, we often realize quite suddenly, in a given situation, that we feel anxious or uncomfortable, pleased or relaxed, and it is apparent that the particular state of feeling we know then has not begun on the moment of knowing but

---

[71] I say "advocates" of this view; however, it is possible (although unlikely) that I am the only advocate of this view. I had thought that my view was 'panexperientialism' until I realised that panexperientialists are part of the *awareness as 'what-it-is-likeness'* view.

rather sometime before. Neither the feeling state nor the emotion that led to it have been "in consciousness," and yet they have been unfolding as biological processes.

Damasio (2000, p.36)

Consider the phenomenon known as hypnotic analgesia in which hypnosis blocks a patient's access to pain, say from an arm in cold water or from the dentist's drill. Pain must be P-conscious, it might be said, but access is blocked by the hypnosis, so perhaps this is P- without A-consciousness? But what reason is there to think that there is any pain at all in cases of hypnotic analgesia? One reason is that there are the normal psychophysiological indications that would be expected for pain...Another (flakier) indication is that reports of the pain apparently can be elicited by Hilgard's "hidden observer" technique in which the hypnotist tries to make contact with a "hidden part" of the person who knows about the pain. The hidden observer often describes the pain as excruciating[72]

Block (1997, pp.405-6)

---

[72] Block, quite confusingly, uses the term 'P-consciousness' to refer to 'feelings states', states which he takes to be wholly devoid of awareness; there is no 'awareness of', no 'conscious of', in P-conscious states, yet they are referred to as conscious states (P-conscious states).Block use of the term 'A-consciousness' can be thought of as referring to awareness; to say the patient's access to pain is blocked is to say that they are not aware of the pain.

6.2.4 *The existence of 'what-it-is-likeness' without awareness*

We have considered various reasons why one might assert that 'what-it-is-likeness' can exist without awareness. One can believe that awareness is a type of higher-order monitoring, one can believe in hypnotic analgesia, one can agree with Damasio's reasoning, and one can believe that most of the universe is 'what-it-is-likeness' but not awareness. Of course, it is also possible to believe that the universe is pervaded by 'what-it-is-likeness', that awareness is a type of higher-order monitoring, that in hypnotic analgesia there is 'what-it-is-likeness' without awareness, and that it is common for there to be states of 'what-it-is-likeness' within one that one is not aware of.

One can think of Damasio's findings, the 'higher-order monitoring' paradigm and hypnotic analgesia as creating a sharp distinction between the two phenomena – *'what-it-is-likeness' can exist without awareness*. Then, from the evolutionary perspective, the 'panwhat-it-is-likeness' advocate takes this insight further and claims that *'what-it-is-likeness' can exist without awareness throughout the universe;* and that only certain parts of the universe have awareness.

6.3 *Awareness without 'what-it-is-likeness'*

In *Section 6.2* we explored the view that states of 'what-it-is-likeness' are not awareness-involving. According to this view states of awareness are directed at states of 'what-it-is-likeness' and take them as their 'content'; this entails that states of 'what-it-is-likeness' can exist in the absence of awareness. In this section our concern is the possibility that states of awareness *themselves* are not 'what-it-is-likeness'-involving.

A state of awareness is a state of the universe which is aware of a state of the universe. If a state of awareness solely becomes aware of a thought such as *two plus two equals four,* then, on the view of thoughts that I have defended, there is no 'what-it-is-likeness' *in the thought.* Is there any 'what-it-is-likeness' *in the awareness?* We seemingly have to conclude that there isn't. For, if one believes that when one becomes aware of the thought: *two plus two equals four,* that there is no 'what-it-is-likeness', then one must believe that there is no 'what-it-is-likeness' in either the thought or the becoming aware of the thought. If there is no 'what-it-is-likeness' to think, but there is 'what-it-is-likeness' to become aware of a thought, then how could one rationally believe that there is no 'what-it-is-likeness' to think? In other words, how could one possibly know if the 'what-it-is-likeness' was a 'what-it-is-likeness' to become aware rather than a 'what-it-is-likeness' to think? The claim that there is no 'what-it-is-likeness' to think *arises from* the belief that there is no 'what-it-is-likeness' to become aware of a thought.

It is also seemingly possible that a state of awareness could take itself as its own content and thereby solely become aware of it*self* at the exclusion of anything external. If you were in a rare state of 'peaceful stillness' – perhaps you were meditating – you might have thought that you were aware in the absence of 'what-it-is-likeness'; perhaps you were just aware of nothing except your awareness.

Why might one believe that awareness itself is 'what-it-is-likeness'-involving? Recall that according to the panwhat-it-is-likeness view the universe is pervaded with 'what-it-is-likeness'. Given this one might believe that on this view awareness is necessarily 'what-it-is-likeness'-involving. However, this isn't actually so due to the universe being a layered (i.e. nested) entity. To say that the entire universe is pervaded with 'what-it-is-likeness' is just to say that one could not find a part of the universe that is wholly devoid of 'what-it-is-likeness'. Given the layered nature of the universe it is possible for there to be both no parts of the universe which are wholly devoid of 'what-it-is-likeness' and for there to be particular 'upper' layers (such as thought and awareness) which are wholly devoid of 'what-it-is-likeness'. So, panwhat-it-is-likeness is compatible with the view that awareness is wholly devoid of 'what-it-is-likeness'. The potential worry that one might have with this is whether it is possible for that which is pervaded with 'what-it-is-likeness' to evolve – when it forms a particular arrangement – an upper layer which is wholly devoid of 'what-it-is-likeness'. However, as with the converse worry – 'what-it-is-likeness' arising out of that which is wholly devoid of 'what-it-is-likeness' – this is clearly not impossible.

What are we to conclude? The most plausible conclusion to reach is that awareness is itself wholly devoid of 'what-it-is-likeness'.[73] If this is so, then we have further grounds for believing there to be a sharp distinction between awareness and 'what-it-is-likeness'. However, if one believes that awareness *is actually* 'what-it-is-likeness'-involving, then the conclusion of *Section 6.2,* that 'what-it-is-likeness' can exist without awareness, still applies.

---

[73] This conclusion will be buttressed when we further consider the phenomenon of awareness in *Sections 6.8* and *6.9.*

6.4 *Perception and awareness*

In *Section 5.2* we considered the human senses and I claimed that the five traditional human senses form two very distinct groups (two senses and three non-senses). The two human senses – seeing and hearing – are clearly engaged in perception; they have what I am calling 'sensory perception'. I claimed that the three 'non-senses' are part of the family of 'feeling states' and I suggested that 'feeling states' are 'particle perceptions'. The purpose of this section is to consider in more detail the nature of perception and specifically the question of whether 'particle perceptions' and 'sensory perceptions' can exist in the absence of awareness.

As we have seen many times already, there are typically 'thick' and 'thin' conceptualisations of phenomena. As we discussed in *Section 4.2.4* the 'thick' conceptualisation of perception entails that perception is awareness-entailing and/or concept-entailing. The 'thick' conceptualisation generally leads to the conclusion that there is a great chasm between humans and their surroundings; this is because it typically leads to the belief that the overwhelming majority of human surroundings are unable to perceive (for instance, most people would reject the idea that a stone has concepts). Contrarily, the 'thin' conceptualisation allows for the possibility that all of human surroundings perceive without awareness or concepts. One will see a regular pattern emerging, as the 'thick' conceptualisation – the conceptualisation which entails the greatest possible chasm – is the dominant contemporary view of perception. A standard contemporary definition of the word 'perceive' is (Oxford English Dictionary, 2009):

To apprehend through one of the senses, esp. sight; to become aware of by seeing, hearing, etc.; to see; to detect.

This definition implies that there is a singular state of the universe which is both perception-involving and awareness-involving. According to this view:

> *Perception is an awareness-involving state of the universe*

Let us turn to the 'thin' conceptualisation. Recall that in *Section 6.2* we considered the possibility that 'what-it-is-likeness' can exist without awareness because awareness is a state which takes other states as its 'content'. Such a 'taking' is the basis of the 'thin' view of perception. According to this view states of perception exist regardless of whether or not awareness takes them as its 'content'. This means that:

> *Perception is a state of the universe that registers information about another state of the universe*

Let us explore how this 'thin' view relates to 'particle perception' and 'sensory perception'. According to this view one's two senses are seeing and hearing continuously, and sometimes one contains states of awareness which take these 'sensory perceptions' as their 'content'. When it comes to 'particle perception' the idea is that all 'feeling states' are states of perception without awareness. So, imagine that one is punched in the leg; this event involves particles in one's leg interacting with a fist. The interaction between the particles in the fist and the particles in one's leg is an interaction which is a meeting of 'feeling states'. The 'feeling states' in the fist interact with the 'feeling states' in the leg and new 'feeling states' are formed in the 'fist-leg' by this interaction. If the particles involved in this interaction register information about each other when they interact then they will perceive each other, and it certainly seems to be the case that information is registered by the respective particles as their states of feeling have been changed due to this interaction. So, 'feeling states' and 'particle perception' are two sides of the same coin. Feeling=particle perception. Particle perception=feeling.

From the panwhat-it-is-likeness perspective the entire universe is pervaded by 'feeling perceptions' – every time two parts of the universe interact this is a 'feeling perception' event.[74] These 'feeling perceptions' are states of the universe which a state of awareness can take as its 'content'. Recall that it is conceivable that a state of awareness can take itself as its own 'content'; contrarily, a state of perception has to be a perception of something else.

We saw in *Section 6.2* that the idea that 'feeling states' can exist in the absence of awareness has numerous supporters. What this means is that if 'feeling states' *are* 'particle percep-tions' then, on this view, 'particle perceptions' exist in the absence of awareness. In the next section our focus is on the issue of whether 'sensory perceptions' can exist in the absence of awareness. We will see that there is plenty of evidence which

---

[74] In *Section 6.5* I claim that the entire universe is in a continuous state of vibration. If this is right it means that every part of the universe will continuously be interacting with its surroundings. In turn, this means that every part of the universe will be a new 'feeling perception' event.

suggests that this is possible, and that the 'thin' view of perception is therefore correct.

6.4.1 *'Sensory perception' without awareness*

The aim of this section is to explore and defend the claim that states of 'sensory perception' – hearing and seeing – can exist in humans in the absence of awareness. This is important because if such states exist then perception and awareness are clearly two different phenomena. Let us consider some of the recent evidence from neuroscience and neuropsychology:

> Twenty (or even ten) years ago a researcher arguing for the existence of subliminal effects was on the fringe of the discipline, on the outside looking in. Now a researcher arguing against the existence of subliminal effects is in that position, while the advocate sits squarely within the mainstream.
>
> (Bornstein, 1992, p.4)

> [A development] that is unique to this age, is the virtual epidemic of dissociations discovered by neuropsychologists whereby residual processing occurs in the absence of acknowledged awareness...blindsight, blind-touch, 'deaf hearing', prosopagnosia and other forms of agnosia, dyslexia, unilateral neglect, and aphasia.
>
> (Weiskrantz, 1992, p.2)

> while some physiological processes which result from sensory stimulation with light or sound may give rise to

awareness of the stimulus, such phenomenal representation is neither a necessary consequence of effective
stimulation, nor a necessary prelude to an overt
response.

(Dixon, 1971, p.2)

Given this evidence for perception without awareness how
should we think about the operations of our two senses? We can
think of them as being engaged in a process of *continuous*
perception of their surroundings; there is no 'off switch'.[75] So,
when the universe arranges itself in such a way as to form a
visual perceptual apparatus and an auditory perceptual apparatus within humans, then these structures perceive until the
moment that the arrangements cease to exist. Whether or not
these perceptions enter awareness is an entirely separate issue.
In support of this view neuroscientist Rodolfo Llinas (2002,
p.118) claims that:

If, while you are awake, someone whispers to you that
there is a bee in your hair, you will most likely do something about it. If, on the other hand, you are asleep when
they whisper, you most likely won't. If this same scenario
of comparisons were under experimental conditions
where it was possible to monitor the flow of auditory
information from your ear into your brain, we would see
that this sensory signal is transduced peripherally, **in**

---

[75] If one agrees with me that arrangements of the universe such as 'video
cameras' are senses, then one will believe that human-created senses *do* have
'off switches' (these senses were designed to be switched on and off) .

full regalia, in both circumstances (waking and sleep). Why don't you hear it when asleep? ...The internal context of the brain during sleep is one that does not grant significance to the meaning of those whispered words or much of any auditory information save for the very loud.

The fact that the auditory apparatus processes the sound "in full regalia" obviously means that the sound was perceived. The reason for the lack of response when asleep seems to be simply that one was asleep! In other words, there was no awareness of the perception; the "internal context of the brain during sleep" is one in which there is no awareness of the perceptions of the auditory and visual apparatus. If there is a very loud sound perceived by the auditory apparatus then this can cause one to stop sleeping/regain awareness. If at the moment of stopping/regaining the sound is still being perceived by the auditory apparatus then one will become aware of it and will realise that one was woken by a loud sound; however, if it was only a momentary sound then one will wake with a startle and will be in a confused state as to why one woke.

Much of the recent scientific evidence for perception without awareness has come from investigating brain damage. Lawrence Weiskrantz (1999, p.8) asserts that: "in virtually all of the major cognitive categories that are disturbed by brain damage, there can be remarkably preserved functioning without the patients themselves being aware of this". One of the most well known examples of such dissociation is the phenomenon of 'blindsight' which occurs when visual stimuli can be discriminated without being 'seen'. Humans with 'blindsight' claim that they are not aware of objects which are immediately in front of them. However, when they are persuaded to guess about whether the objects are moving in a certain direction they are able to do so correctly the vast majority of the time. When these humans are informed that they have actually perceived the correct movements of the objects in the vast majority of cases, they are greatly surprised. The phenomenon of 'blindsight' strongly suggests that it is possible for humans to detect and localise visual stimuli without awareness.

Melvyn Goodale and David Milner have recently provided an account of the visual system which explains both blindsight and other cases of visual perception without awareness. They argue that there are two visual perceptual systems – the 'dorsal action system' which can never enter awareness, and the 'ventral perceptual system' which can potentially enter awareness. Goodale and Milner (2004, p.30, pp.47-8) assert that:

Rather than evolving some kind of general-purpose visual system that does everything, the brain has opted for two quite separate visual systems: one that guides our actions and another, quite separate system, that handles our perception.

the visuomotor networks [of the dorsal action stream] no more need conscious representations of the world than does an industrial robot. The primary role of perceptual representations [of the ventral perceptual system] is not in the *execution* of actions but rather in helping the person or animal to arrive at a decision to act in a particular way.

After presenting an array of scientific evidence to support their view Goodale and Milner (2004, p.55) conclude that: "there remains a whole realm of visual processing that we can never experience or reflect on. We are certainly aware of the actions that these visuomotor systems control, but we have no direct experience of the visual information they use."

So, this is further clear support for the existence of perception without awareness. Information is obtained from the surroundings and is used to control action, but it can never enter awareness. Goodale and Milner, perhaps surprisingly, deny that the activity of the *dorsal action stream* entails 'perception'; this is simply because they adopt the standard contemporary definition of 'perception' – the 'thick' conceptualisation – according to which perception necessarily entails awareness. Nevertheless, Goodale and Milner (2004, p.114) do argue that perception without awareness occurs in the *ventral perceptual system* as: "the visual computations underlying unconscious perception seem to be identical to those underlying conscious perception: it's just that they don't make it into awareness."

Let us conclude. We have seen that there is a plethora of evidence which supports the idea that perception can exist without awareness. This means that the 'thin' view of perception appears to be correct. Perception and awareness are two different phenomena.

6.5 *Perception and 'what-it-is-likeness'*

There are two relationships which we need to consider – the relationship between 'sensory perception' and 'what-it-is-likeness', and the relationship between 'particle perception' and 'what-it-is-likeness'. We have, in large part, covered this relationship in previous sections. In *Section 6.4* we concluded that 'feeling states' and 'particle perception' are two sides of the same coin. Feeling=particle perception. Particle perception=feeling.

In *Chapter Five* we concluded that the 'what-it-is-likeness' of the two human senses is located in the universe independently of whether or not it is perceived; the universe is pervaded with both colours and sounds. What this means is that all 'sensory perceptions' will be 'what-it-is-likeness'-involving. When the visual perceptual apparatus perceives its surroundings it will inevitably perceive 'what-it-is-likeness'; when the auditory perceptual apparatus perceives its surroundings it will inevitably perceive 'what-it-is-likeness'. This is the case regardless of whether or not there is awareness of these 'sensory perceptions'.

Something we haven't yet considered is whether or not there can be 'what-it-is-likeness' *without* perception. This is an interesting question. If one believes that thought is 'what-it-is-likeness'-involving then one will believe that there can be 'what-it-is-likeness' without perception; for, thought is not perception-involving (to think *two plus two equals four* does not entail any interaction with the surroundings surrounding the thought). However, I have already defended the idea that thought *is not* 'what-it-is-likeness'-involving; recall that there are only two groups of 'what-it-is-likeness' – 'feeling states' and 'sensory what-it-is-likeness'. There cannot be 'feeling states' without perception, because 'feeling states' *are* perceptions. Can there be 'sensory what-it-is-likeness' without perception? Recall that 'sensory what-it-is-likeness' is located in the surroundings of the perceptual apparatus and not in the perceptual apparatus. So, the answer to our question depends on whether or not the 'sensory what-it-is-likeness' that pervades the universe is perception-involving. In other words: Are colours and sounds perceptions?

This might sound like a slightly odd question; for, a human normally thinks of colours and sounds as things *that they perceive* rather than as things *that perceive*. However, recall that in *Section 5.2* we encountered the 'pansensism' paradigm and I claimed that this view should be interpreted as entailing that 'particle perceptions' pervade the universe ('panperceptualism'). I have also defended the view that the entire universe is pervaded with colours and sounds. Furthermore, we have seen that sounds are vibrations, and that there is seemingly an 'infinite manifold' of colours in the universe, most of which humans are unable to perceive. It is also the orthodox view that the entire universe is in a continuous state of vibration; the fundamental constituents of the universe are never in a state of 'motionless rest'. According to physicist Brian Greene (2000, p.146): "The universe – being composed of an enormous number of these vibrating strings – is akin to a cosmic symphony."

If we combine all of this we have a particular world view. Every vibration in the universe is a sound, every vibration in the universe is a colour, every vibration in the universe is a feeling (this would be the panwhat-it-is-likeness view of a vibrating universe), and every vibration in the universe is a perception. In short, the universe is a collection of vibrations, and every vibration is a perception-involving change in sound/colour/feeling. Let me put this slightly differently. Consider a single vibration – according to this view this vibration is *simultaneously* a particular sound, a particular colour and a particular feeling. If the association is tight enough then it would be the case that a particular colour = a particular sound = a particular feeling = a particular perception. If this is so, then, in effect, colours are perceptions (as well as sounds and feelings).[76] If this sounds implausible recall that *it is* very plausible that there are 'infinitely' more colours and sounds in the universe than humans can possibly perceive.[77] In other words, this is not a claim about the universe that is accessed by the

---

[76] This is why the Martian Objection is not applicable. Soundwaves are a human conceptualisation. In the 'universe in-itself' there are vibrations of sound=colour=feeling=perception. When a perceiver (human or Martian) evolves a perceptual connection to their surroundings they are evolving a connection to *a particular aspect* of these vibrations. This view entails that a human and a Martian could connect to the very same vibration and whilst the human connects to colour the Martian connects to sound.

[77] Recall that in *Section 5.4.6* I defended the view that the universe is an 'infinite manifold of physical colours'. This view entails that where Tom perceives green, Jane can perceive red; this implies that more than one colour can simultaneously exist in the same place. However, this is not necessarily so. For, whilst Tom can perceive a particular tomato to be green at the same time as Jane perceives this tomato to be red, the part of the 'blobject' that is the tomato can simultaneously contain a multitude of *sound=green=feeling=perception* vibrations and *sound=red=feeling=perception* vibrations. Tom's visual perceptual apparatus connects to the former vibrations; Jane's (different) visual perceptual apparatus connects to the latter vibrations. Another possibility is that the connection between *sound=colour=feeling=perception* is not as tight as this; whilst every vibration is sound, colour, feeling and perception, a *particular* vibration could be *multiple* colours, *multiple* sounds, and *multiple* feelings. On this latter possibility the smallest constituents of the universe are simultaneously multiply coloured; on the former possibility the smallest constituents of the universe are determinately coloured, but when two perceivers perceive a particular part of the 'blobject' they can simultaneously perceive different colours.

human perceptual apparatus – a segment which has a particular range of accessible colours and sounds; it is a claim which concerns the 'universe in-itself.' If this is right then there cannot be 'what-it-is-likeness' without perception; for, 'feeling states', 'colours' and 'sounds' are all perception-involving.

6.6 *Awareness, sleep and dreaming*

In *Section 6.4.1* we considered how the auditory perceptual apparatus can perceive sounds "in full regalia" whilst a human is asleep; if the human was awake they would respond to these auditory perceptions, but because they are asleep they do not respond. I claimed that the reason for the lack of response whilst asleep was that there was no awareness of the perception. Indeed, you probably think that this is obvious. Awareness is seemingly a state that doesn't exist in humans when they are sleep, and is a state which (at least typically) exists in humans when they are not asleep.[78] Could it be that there is a connection of potential importance here? Could it be that there is a fundamental link between sleep and awareness in humans? If so, then it is possible that there could be a more general connection – sleep and awareness being reciprocal states of a particular arrangement of the universe. To put this slightly differently, the arrangements of the universe which have evolved to sleep could also be the arrangements which have evolved to be aware. To believe that the way that the universe has evolved means that only those parts of the universe which sleep have the capacity for awareness is not to make the more general claim that *it is impossible for arrangements of the universe that do not sleep to become aware.* Contrarily, it is just to make a claim about the

---

[78] The purpose of this section is to explore whether this seeming is an actuality.

way that the universe has actually evolved. One can also accept that humans, through their technological expertise, can bring forth arrangements of the universe which are aware but have no need for sleep.

It is seemingly obvious that the way that the universe has evolved means that *only some parts* of the universe sleep. It is also very plausible to believe that the way that the universe has evolved means that *only some parts* of the universe contain states of awareness. Additionally, there also seems to be an *obvious connection* between sleep and awareness. So, the possibility that the arrangements of the universe that have the capacity for awareness *are also* the arrangements of the universe which sleep should be seriously considered.

There are no obvious answers to the question of why the universe evolved particular arrangements which sleep. One possibility is that generating awareness requires a lot of energy and is thus not a state which can easily be continuously maintained. If this is so, then periods of rest – of absence of awareness – would clearly be required; and these periods would be the periods we call 'sleep'. If this is right, or anywhere near the mark (or possibly nowhere near the mark), then the arrangements of the universe which contain states of awareness will be, more or less, those parts which sleep. This implies that more or less all animals sometimes contain states of awareness. Let us consider the following assertions from David Papineau and John Searle:

> Sometimes consciousness is explained as the difference between being awake and being asleep. But this is not quite right. Dreams are conscious too...Consciousness is

what we lose when we fall into a dreamless sleep or undergo a total anaesthetic.

(Papineau, 2005, pp.4-5)

"consciousness" refers to those states of sentience and awareness that typically begin when we awake from a dreamless sleep and continue until we go to sleep again, or fall into a coma or die or otherwise become "unconscious." Dreams are a form of consciousness, though of course quite different from full waking states. Consciousness so defined switches off and on.

(Searle, 1997, p.5)

The first thing to note is that both Papineau and Searle believe that there is an obvious connection between sleep and awareness – awareness is a state of the universe which can vanish when the part of the universe that instantiates it falls asleep. According to this view awareness is a discontinuous phenomenon (it pops into and out of existence) and the arrangements of the universe that sleep are arrangements of the universe which have the capacity for awareness.

The second thing to note is that both Papineau and Searle have a 'thick' conceptualisation according to which awareness='what-it-is-likeness'; this is what they implicitly assume when they use the word 'consciousness'. They both claim that the 'popping boundary' is not simply between sleep and awareness, rather they claim that it is located between states of *dreamless* sleep and awareness. Searle claims that: "Dreams are a form of consciousness". Why do Papineau and Searle believe that dreams are 'conscious'? There are two main theories concerning the nature of dreams – either one is aware of them when they are occurring or one becomes aware of them after they have occurred; Papineau and Searle clearly have the first view. One motivating factor underlying this choice of view might be that they believe that there is a 'what-it-is-likeness' to have dreams; if one believes this and also believes that awareness='what-it-is-likeness', then one has to conclude that dreams are 'conscious'. If one adopts the second view of dreams then one can simply assert that the location of the 'popping boundary' is between sleep (which includes dreams) and awareness.[79]

The situation is complicated by the fact that there appear to be two different types of dreams – lucid dreams and non-lucid dreams. Stephen LaBerge (2005, p.141) explains the phenomenon of lucid dreaming as follows:

> what happens when you become lucid? Essentially, you become explicitly aware of a particular important fact – that you are dreaming...if you were sleeping in a sleep lab with electrodes to record your eye-movements, you could mark the moment when you became lucid by, for example, looking to the left, right, left, and right in the dream. Then let's say you flew about your dream and then woke up a few minutes later and reported your dream. The polygraph would in fact show the eye-movement signal just when you reported.

---

[79] It is possible for perceptions such as sounds to be incorporated into dreams. Recall from *Section 6.4.1* that perception can exist without awareness, so this phenomenon does not provide any support for the view that dreams are states of awareness.

What transition is occurring when one 'becomes lucid'? It is the transition from lacking the attribute of awareness to attaining the attribute of awareness. What are the implications of this? The obvious implication is that the vast majority of dreams (non-lucid dreams) are wholly devoid of awareness – they are on the sleep/non-aware side of the 'popping boundary', whilst the 'lucid dreams' do contain awareness – they are on the aware side of the 'popping boundary'. If this is right, it means that anyone who makes a broad-brush statement such as "dreams are conscious" is strictly wrong – as the vast majority of dreams (non-lucid dreams) would be wholly devoid of awareness.

Despite the various distinctions in the realm of dreams which I have been exploring – each of which leads to a slightly different location for the 'popping boundary' between awareness and non-awareness – all of these positions are united by their belief in the existence of the 'popping boundary'. Does the 'popping' actually occur? Richard Unger (2006, p.46) considers and swiftly rejects Descartes view that there is no such 'popping':

> Descartes held, against common sense (and incorrectly) that there never were any periods of his sleep, in any night or day, when he wasn't conscious; it *only appeared that way.*

According to the account of perception, awareness and 'what-it-is-likeness' that I have been proposing in this chapter Descartes would be correct to assert that there were never any periods when he wasn't thinking, perceiving or feeling.

However, he would be wrong to assert that there are never any periods when he is devoid of awareness. Let us side with common sense, and with Unger, Searle, and Papineau, and conclude that awareness is a discontinuous phenomenon which only exists when a particular arrangement of the universe 'pops' out of the abyss of unawareness.

6.7 *The relationship between awareness, perception and 'what-it-is-likeness'*

At the start of the chapter I outlined the dominant *awareness as 'what-it-is-likeness'* paradigm according to which awareness and 'what-it-is-likeness' are one and the same state of the universe. We saw that this view is rejected by the 'higher-order monitoring' theorists, the panwhat-it-is-likeness advocates, and by others, because *'what-it-is-likeness' can exist without awareness.*

We explored the idea that *perception can exist without awareness* and concluded that this is possible. Furthermore, we also concluded that perception is necessarily 'what-it-is-likeness'-involving. The existence of perception without awareness, combined with perception necessarily being 'what-it-is-likeness'-involving, means that *'what-it-is-likeness' can exist without awareness.* This is the same conclusion that we reached in the previous paragraph. It is obviously encouraging that we have two different routes to the same conclusion.

We also explored the possibility that all 'what-it-is-likeness' is perception-involving and concluded that this is a possibility. If one combines panperceptualism with the idea that the universe is a collection of vibrations, then every vibration could be a perception-involving change in sound/colour/feeling; this means that every instance of 'what-it-is-likeness' is perception-involving. We also concluded that it is likely that awareness can exist without 'what-it-is-likeness'.

Finally, we explored the links between these three phenomena and the states of sleep and dreaming. The common sense idea that awareness is a discontinuous state was explored and defended; awareness is a state of the universe that 'pops' into and out of existence. Furthermore, I contended that there is an intimate connection between awareness and sleep – the arrangements of the universe that sleep are, more or less, those that have the capacity for awareness. It is not absolutely clear on which side of the 'popping' boundary between sleep and awareness that normal dreaming and lucid dreaming should be placed. However, I suggested that the answer to this question is not of great importance; what is important is that the 'popping' boundary exists. The existence of perceptions and 'what-it-is-likeness' without awareness implies that there can be perceptions and 'what-it-is-likeness' in a human when they are asleep, and we concluded that this is indeed the case.

In short, perception and 'what-it-is-likeness' appear to be a tightly coupled, if not inseparable, 'pair' of phenomena. However, awareness appears to be a singular state which is wholly separable from both perception and 'what-it-is-likeness' (or perception='what-it-is-likeness'). This conclusion is clearly potentially of great importance when we consider the relationship between humans and their surroundings from an evolutionary perspective – perhaps awareness evolved in humans, but perception='what-it-is-likeness' pervades the surroundings which evolved humans. Let us further consider the phenomenon of awareness in order to see whether this is plausible.

6.8 Are there levels of consciousness and altered states of consciousness?

There is widespread talk of different 'levels' of consciousness and of 'altered states' of consciousness. As we saw in *Section 6.6* Searle even talks about different 'forms' of consciousness. If one has a very 'thick' conceptualisation of consciousness then such talk seems to make some kind of sense. For, if a large enough range of states of the universe are brought under the umbrella term 'consciousness' then it seems almost inevitable that it will be appropriate to talk about different levels, different forms and altered states. However, if one has a 'thin' conceptualisation of consciousness – awareness – does it make sense to talk about different levels of awareness and/or altered states of awareness? In other words, if one asserts that one is in an 'altered state of awareness', or that there are different 'levels' of awareness, is one talking some kind of nonsense? In *Section 6.8.1* I outline the widespread belief that there are such things as the aforementioned 'levels' and 'altered states'. In *Section 6.8.2* I present an alternative account according to which there are no 'levels' and 'altered states'.

6.8.1 The 'levels' and 'altered states' paradigm

Talk of the existence of different 'levels of consciousness' and 'altered states of consciousness' is very widespread. What exactly do humans mean when they utter these phrases? There seems to be two different ways in which the terms can be used; I will refer to these as the 'spatial sense' and the 'temporal sense'. In the 'spatial sense' the claim that that there exist in the universe different 'levels of consciousness' is the claim that it is possible to identify two different areas of the universe which possess the attribute of consciousness and one of them will possess a 'higher level' of it than the other. Similarly, in the 'spatial sense', the claim that that there exists in the universe 'altered states of consciousness' is the claim that it is possible to identify two different areas of the universe which possess the attribute and one of them will possess this state in a way that is somehow 'altered' from the other.

What about the 'temporal sense'? In the 'temporal sense' the claim that there are different 'levels of consciousness' is the claim that one particular spatially located area of consciousness can switch from a low level of consciousness at t_1 to a high level of consciousness at t_2. Similarly, in the 'temporal sense' the claim that that there are 'altered states of consciousness' is the claim that one particular spatially located area of consciousness can switch from a particular type of consciousness at t_1 to an 'altered' type of consciousness at t_2.

So, the 'spatial sense' concerns the existence of states in different parts of the universe; for example, a human and a fish. Contrarily, the 'temporal sense' concerns the states that exist in one part of the universe; for example, a particular human.

6.8.1.1 *The 'temporal sense'*

What could it possibly mean to say that there are 'altered states of consciousness' or 'different levels of consciousness' in a particular human at different times? If one has a 'thick' conceptualisation of consciousness then these terms can be used as follows. One is exceedingly drunk and asserts "I am in an altered state of consciousness". One is feeling very sleepy and inattentive and asserts: "I am at a low level of consciousness". Let us consider the following statement from Charles T. Tart (1990, pp.1-2):

> there is a multitude of philosophical and semantic problems in defining just what "normal" consciousness and "altered" states of consciousness are, yet at this instant I have not the slightest doubt that I am in my normal state of consciousness. Yet there have been a number of occasions in my life when I have not had the slightest difficulty in realizing that I was in an altered state of consciousness.

> An altered state of consciousness for a given individual is one in which he clearly feels a *qualitative* shift in his pattern of mental functioning, that is, he feels not just a quantitative shift (more or less alert, more or less visual imagery, sharper or duller, etc), but also that some quality or qualities of his mental processes are *different*.

So, Tart is certain that he has been in 'altered states of consciousness' despite not really understanding what this assertion means. Given what he says it is clear that what he takes the assertion to mean is that in some instances he has become aware of "qualitative" states of 'what-it-is-likeness' that are not normal for him. So, the phrase 'altered states of consciousness' effectively means an abnormal state of 'what-it-is-likeness'. If one has a 'thick' conceptualisation of consciousness according to which awareness and 'what-it-is-likeness' do not come apart then any alteration in 'what-it-is-likeness' is an alteration in consciousness. However, if one has a 'thin' view of consciousness according to which awareness is a distinct state from 'what-it-is-likeness', then one should strictly say that one is an 'altered state of what-it-is-likeness' (one is *not* in an altered state of awareness!).

There are undoubtedly times when a human feels particularly alert and their 'sensory perceptions' are 'razor sharp'. These situations compare to those situations when this human feels drowsy and not alert. Again, if one has a 'thick' conceptualisation of consciousness then to say that these situations are different 'levels' of consciousness makes sense. However, if one has the 'thin' view according to which awareness and its contents are distinct states, then it would be incorrect to say that the 'level' of awareness *itself* has changed.

6.8.1.2 The 'spatial sense'

Let us consider the possibility that there are two different parts of the universe which instantiate consciousness in 'altered ways' or at 'different levels'. The 'spatial sense' is typically associated with the belief that there are 'different levels' of consciousness so this will be our focus. It is widely claimed that the human species has a higher level of consciousness than other species of animals; it is even often asserted that there is a 'sliding scale of consciousness' throughout the animal kingdom. Furthermore, it is also frequently claimed that some humans have a higher level of consciousness than others; this is particularly the case when certain humans are hypothesised to have become 'spiritually enlightened'. These widespread beliefs are aptly described by R. M. Bucke (1905, pp.1-2):

> To make the matter clear it must be understood that there are three forms or grades of consciousness. (1) Simple Consciousness, which is possessed by say the

upper half of the animal kingdom. By means of this faculty a dog or a horse is just as conscious of the things about him as a man is; he is also conscious of his own limbs and body and he knows that these are a part of himself. (2) Over and above that Simple Consciousness, which is possessed by man as by animals, man has another which is called Self Consciousness. By virtue of this faculty man is not only conscious of trees, rocks, waters, his own limbs and body, but he becomes conscious of himself as a distinct entity apart from all the rest of the universe... (3) Cosmic Consciousness is a third form which is as far above Self Consciousness as is that above Simple Consciousness. With this form, of course, both simple and self consciousness persist (as simple consciousness persists when self consciousness is acquired), but added to them is... a consciousness of the cosmos, that is, of the life and order of the universe.

According to this account there are three different levels of consciousness. This means that if one were to identify all of the states of the universe that are currently states of consciousness then one could carve them up into three groups: firstly, the states of simple consciousness; secondly, the states of self consciousness; thirdly, the states of cosmic consciousness. If one has a 'thick' conceptualisation of consciousness then these varying levels of consciousness seem to exist. However, if one has a 'thin' conceptualisation of consciousness, then one will not believe that the various states of the universe described by Bucke are different levels of awareness. Contrarily, one will believe that Bucke's *Stage 1* is an account of awareness. In tandem with this one will believe that the progression that Bucke describes is a progression of the things that awareness becomes aware of; it is the progression of thought and under-standing.

6.8.2 *The 'variable contents' paradigm*

We have been considering the *'levels' and 'altered states' paradigm;* the alternative to this paradigm is the *'variable contents' paradigm.* Whilst the intelligibility of the former depends on a 'thick' conceptualisation of consciousness, the intelligibility of the latter is dependent on a 'thin' conceptualisation of consciousness.

If the 'thin' conceptualisation of consciousness is correct then when humans use the phrases 'altered states of conscious-ness' and 'different levels of consciousness' they are using words that are strictly inappropriate. They could more accurately say something along the lines of: *this is an abnormal state of 'what-it-is-likeness' which I am aware of;* or, *I am aware that I am in a drowsy state at the moment.* According to the 'thin' view of consciousness, which is the view I have been defending, awareness is not itself something which can 'alter' or ascend to higher 'levels'. The states of the universe that are states of awareness simply have different 'contents'. So, if one is in an 'altered state' of consciousness, then the states of the world which have 'altered' are the 'contents' of awareness, not aware-ness itself. According to this view awareness is 'level-less' and has variable contents.

6.8.3 *Conclusions*

From an evolutionary perspective on the relationship between humans and their surroundings is it important whether one believes in the *'variable contents' paradigm* or the *'levels' and 'altered states' paradigm?* If one has a 'thin' conceptualisation of consciousness and therefore believes that awareness is 'level-less' and has 'variable contents', then there is a good chance that one will believe that most of the universe is devoid of awareness. However, if one believes that there can be different levels of awareness, then one can easily find oneself on a slippery slope towards attributing awareness to things such as atoms, and calculators, on the basis that *it is only a very low level* of awareness that these entities have. William Lycan (1996, p.43) claims that:

> notebook computers, etc., are conscious to a very small degree and lie on a long continuum that has us at, or with an eye toward the future, toward, the other end.[80]

If one believes in the 'variable contents' paradigm then one could agree with Lycan that notebook computers are conscious – that they contain states of awareness. However, it seems more likely that the 'variable contents' advocate will attribute *non-conscious* 'contents' to notebook computers – perhaps 'feeling states'/perceptual states and/or thought states. So, I am suggesting that if one believes that consciousness can exist at varying 'levels' and in 'altered' states, that one is much more likely to draw the seemingly implausible conclusion that notebook computers are aware, or even that the entire universe is aware to some degree. Indeed, this is the conclusion of the

---

[80] It might be worth repeating what I said earlier. I take it that any intelligible use of the word 'conscious' entails that if one says that "X is conscious" one means that "X is aware".

'panawareness' paradigm which we first encountered in *Section 3.5.4.*

6.9 The 40-Hz oscillation theory

Let us recap. In *Section 6.7* we concluded that there are good reasons to believe that awareness is a singular state which is wholly separable from both perception and 'what-it-is-likeness', but that there seems to be a very intimate connection between perception and 'what-it-is-likeness'. In *Section 6.8* we further explored the phenomenon of awareness and I proposed that despite widespread talk about 'altered states of consciousness' and different 'levels of consciousness', that awareness is actually level-less and has variable contents.

Is this level-less view of awareness supported by everything that we know about the human brain? In this section my aim is to show that a dominant neuroscientific theory of consciousness – the 40-Hz oscillation theory – is compatible with the level-less view of awareness. Furthermore, when we attempt to understand which arrangements of the universe are states of awareness, the arrangements proposed by this theory are a very good candidate.

The 40-Hz oscillation theory was first proposed by Francis Crick and Christof Koch (1997, p.277) who claimed that consciousness requires an attentional mechanism:

> This attentional mechanism helps sets of the relevant neurons to fire in a coherent semi-oscillatory way, probably in the 40-70 Hz range, so that a temporal global

271

unity is imposed on neurons in many different parts of the brain.

Scientific studies imply that a necessary condition for awareness is the presence of the two sets of intralaminar nuclei (ILN) and the reticular formation. This has led Bernard J. Baars and Rodolfo Llinas to assert that:

> One hypothesis is that the ILN works with another small nucleus, the reticular nucleus, to generate a regular waveform about forty times per second, which may serve to coordinate and "bind" many specific areas of cortex into a single, conscious experience.
>
> (Baars, 1997, p.31)

> If we posit that the 40-Hz coherent waves are related to consciousness we may conclude that consciousness is a noncontinuous event determined by simultaneity of activity in the thalamocortical system...A 40-Hz oscillation displays a high degree of spatial organization and thus may be a candidate mechanism for the production of the temporal conjunction of rhythmic activity over a large ensemble of neurons.
>
> (Llinas, 2002, p.124)

This theory fits perfectly with the level-less view of awareness. The 40-Hz oscillations would be states of awareness which take other states – such as states of 'what-it-is-likeness' and 'sensory perceptions' – as their 'contents' as the oscillations spread out from their point of origin. However, the proponents of this theory are entrenched within the 'thick' conceptualisation of consciousness, believing that the 40-Hz oscillations bring 'what-it-is-likeness' into the universe. Christof Koch (2005, pp.128-30) states that:

> it's these neurons that have a standing wave between the inferotemporal cortex and prefrontal cortex, and feedback, and then you're conscious–fine, but why does it give rise to the subjective feeling? Right now the answer is 'I don't know'...I don't understand why some neural activity feels like something. I mean I really don't know.

The belief that these oscillatory patterns bring 'what-it-is-likeness' into the universe has led Ned Block (2004, p.210) to object to the theory:

> Francis Crick and Christof Koch (1990) have famously hypothesized that the neural basis of consciousness is to be found in certain phase-locked 40Hz neural oscillations. But how does a 40Hz neural oscillation explain *what it's like* (in Nagel's memorable phrase) to be us?

This objection is entirely understandable and surely correct. Indeed, it is hard to take seriously the idea that a pattern of 40-Hz oscillations could bring into existence the plethora of states of 'what-it-is-likeness' that a human becomes aware of. However, when we modify the 40-Hz oscillation theory, when we bring it together with the level-less view of awareness, then everything falls into place. According to this view 40-Hz oscillations most definitely *do not* bring 'what-it-is-likeness' into existence, they simply bring the awareness of pre-existing 'what-it-is-likeness' into existence.

What does this conjunction of the 40-Hz oscillation theory of consciousness with the level-less view of awareness imply about which arrangements of the universe are awareness? Obviously it implies that wherever there are 40-Hz oscillations there is awareness. So, when humans are wholly devoid of these oscillations, for example, when they are asleep, there will be no awareness in these humans. This is happily concordant with our earlier conclusion that awareness is a discontinuous state and that there is a 'popping' boundary between sleep and awareness. Furthermore, it also follows that numerous species of animals will contain states of awareness, but contra-Lycan notebook computers will not contain states of awareness, and contra-panawareness the entire universe will not be aware.

From the evolutionary perspective we have good reasons to believe that the attribute of awareness distinguishes humans from their surroundings *when humans are aware.* In other words, when humans contain 40-Hz oscillations they are distinguished from the vast majority of their surroundings because they are aware. However, when humans do not contain 40-Hz oscillations then they are not distinguished from their surroundings through the attribute of awareness. Furthermore, given the separation between awareness and 'what-it-is-likeness' we can see the attractiveness of panwhat-it-is-likeness; according to this view the entire universe is pervaded with 'what-it-is-likeness' and it is only in certain parts of the universe

– where 40-Hz oscillations exist – that the awareness of 'what-it-is-likeness' has evolved.[81]

I have claimed that awareness is a discontinuous phenomenon and I have contended that 40-Hz oscillations are a good candidate for the arrangements of the universe that are awareness. Is it possible that arrangements of the universe that are not 40-Hz oscillations are also awareness? This is obviously possible. However, if we accept that 40-Hz oscillations are arrangements of the universe that are states of awareness, then there seems to be little motivation to hypothesise that different arrangements of the universe are also states of awareness.

6.9.1 The 'area of awareness'

Let us consider the transition that occurs from unawareness to awareness in a particular arrangement of the universe. At one moment awareness does not exist; then, in the next moment there is a change in the arrangement of the universe in this area – a particular oscillatory pattern arises and spreads out thereby creating an 'area of awareness'. This 'area of awareness' is constantly in a state of flux; for, oscillations are intrinsically dynamic and will thereby cause the edges of the 'area of aware-ness' to be blurry. The 'area of awareness' is roughly confined to the brain in which it is generated; however, it can take as its

---

[81] Recall that in *Section 4.2.6* we considered the possibility that such oscillations could intelligibly exist in parts of the universe such as a magnet. It is possible that when humans become sufficiently knowledgeable and technological that they will be able to impart 'awareness oscillations' onto many diverse arrangements of the universe.

'contents' states which are non-brain located, such as the 'feeling state' of pain that is located in a 'left foot'.[82]

Areas of the brain which oscillate 'out of synch' due to being damaged will not be incorporated into the 'area of awareness'; this can explain phenomena such as 'blindsight' which we explored earlier. The coming into existence and the going out of existence of these oscillations are a central part of the life of a human. When the oscillation of this oscillation pattern ceases then this typically means that the human in which the pattern was instantiated has also ceased.

---

[82] Given my definition of 'a sense' in *Section 5.2* – which entails that only two of the five traditional human senses are senses – awareness would be a third human sense which *sometimes* exists in humans.

Chapter 7

Human specialness and uniqueness

So far we have considered a number of attributes which are widely believed to create a chasm between humans and their surroundings. For example, human surroundings are 'natural' whilst humans are not natural, human surroundings are non-minded whilst humans have minds, and human surroundings are wholly devoid of 'what-it-is-likeness' whilst humans have 'what-it-is-likeness'. Furthermore, I have also sought to understand why this situation obtains. Imagine that one was told that there was a universe which was created in a Big Bang and that it had been very gradually evolving slightly different arrangements and patterns ever since, and that nothing new had been added to this universe, that that which exists now is that which had always existed. On hearing this story the most plausible thing to believe would be that there were no chasms in this universe – all parts of the universe would be qualitatively similar throughout. However, this is clearly not what one typically believes when one is a human *situated within* such a universe. I have explored why this might be so by exploring the way in which the human perceptual apparatus has evolved to perceive its surroundings in a particular way, and concluded that this constrained way in which perceptions arise is a central factor underpinning the creation of world views which have chasms at their core.

The purpose of this chapter is two-fold. Firstly, following our consideration in *Chapter One* of humans doubting their naturalness, I seek to consider in greater depth why this is so. I claim that the answer lies in the realm of advanced tool use because this generates a 'sense of specialness'. This consideration is to be found in *Section 7.1 – A sense of specialness*. Secondly, I seek to consider all of the attributes which, it seems to me, could plausibly be believed to create a chasm between a human and *all of* their non-human surroundings. One can fruitfully think of this consideration as occurring within the 'static view' – take a human, take a non-human part of the universe, compare them: Does the human have a chasm-creating attribute? This is the subject of *Section 7.2 – Do humans have a unique attribute?*

7.1 A sense of specialness

In our current epoch the dominant world view entails that there is a chasm between humans and their surroundings. Furthermore, the nature of this chasm is that it supposedly elevates humans to a position of superiority – of specialness. This is revealed by a plethora of phenomena such as the biblical assertion that humans have dominion over all other life-forms on the Earth, and the deeply ingrained view of (non-human) animals as human inferiors – if you want to insult a human just call them an 'animal'! It is also revealed in the repeated pattern we have encountered: the dominant conceptualisations of various human attributes are all 'thick' conceptualisations which tend to result in the 'widest' possible conceptual division between humans and their surroundings.

Why does this sense of human specialness – *this sense that the human species is superior to the rest of the Earth (both non-*

human life-forms and the non-living parts) – arise?[83] If one goes far enough back in the life of a human then one will inevitably arrive at the moment when this sense of specialness first arose in that human. It seems reasonable to assume that human infants do not have this sense of human specialness. When human infants explore their surroundings they surely don't have a sense that the human species is somehow special, somehow divided from the other planetary life-forms.

Perhaps it is when a human reaches what one can loosely call a 'broad overview' of their surroundings that this sense of specialness arises. A 'broad overview' of one's surroundings entails a reasonably broad range of knowledge about those surroundings. So, if one knows that there are countries and continents, mountains and oceans, fish and birds, insects and bacteria, planets and stars, boats and cars, aeroplanes and space stations, apes and dolphins, mobile phones and laptop computers, and cats and dogs, then one has a 'broad overview' of one's surroundings.

It seems almost inevitable that when a human first has a 'broad overview' of their surroundings that they will have a sense that the human species is special. Why is this? It is because humans seem to have abilities that other things lack. Let us consider an extreme case in order to make the point: if a human grows up surrounded with designer clothes, a nice

---

[83] I should make it clear that the 'sense of human specialness' is not something that is rationalised – it is something which leads to rationalisations. The 'sense of specialness' arises almost automatically within humans, after it has arisen humans then ask themselves the question: *What is it that makes the human species special?* In other words: *Why isn't the human species just another species of animal?*

house, books, cars, aeroplanes, skyscrapers, mobile phones and computers, then one should not be surprised if a sense of human specialness arises within that human. After all, no other species of animals can be found reading books, flying aeroplanes or texting on their mobile phones.

It is very easy to understand why a human can have *a sense* that their species is special. But is the human species *actually* special? And, if it is, *why* is it special? I am convinced that there is a *sole source* of the sense of human specialness: the advanced tool-using ability of the human species. It is because humans are born into, and grow up in, surroundings which are pervaded with *advanced human tools* that the sense of human specialness arises. If this is right, then in the early stages of the evolution of the human species there would not have been a widespread sense of human specialness. I would suggest that the tools used by hunter-gatherers – bow and arrow, harpoon, atlatl, projectile points – are advanced enough to give rise to a sense of human specialness.

What, exactly, is an advanced tool? I take the following to all be examples of advanced tools: knives, forks, buckets, tee-pees, cups, aeroplanes, harpoons, books, submarines, laptop computers, cars, televisions, pens, bicycles, skyscrapers, space-craft, tables and chairs; some of these advanced tools are clearly more advanced than others. These tools are 'advanced' compared to 'non-advanced' tools such as twigs and stones. 'Non-advanced' tools are still used by humans today and they were the only tools used by early humans. I am proposing that when the human species started to modify its surroundings in more than the simplest of ways that this resulted in a widespread sense of specialness within the human species. So, the source of the widespread sense of human specialness *which exists today* is advanced tool use.

You might not be convinced by this – you might believe that your sense that the human species is special has an alternative source. For instance, you might be thinking: *if all of the advanced tools on the Earth disappeared I would still have a sense that the human species is special.* And, of course, you would be correct! You have grown up in surroundings which are pervaded with advanced human tools and therefore have a sense of human specialness; once this sense of human specialness has arisen it will not vanish simply because all of the advanced tools on the Earth have disappeared.

What we are interested in is whether a human who is born into, and grows up in, surroundings which are wholly devoid of advanced human tools would have a sense that the human species is special – a sense that there is a fundamental division between the human species and all of the other planetary life-forms. On the Earth today it would be hard to find such a human. There are still small groups of humans who live on secluded islands and deep within rainforests but these groups use advanced human tools. There is always a chance that a true 'feral' child will be found – a human who from birth was raised by non-human animals in the wild. I am fairly convinced that such a human would not have a sense that the human species is special.

It seems that there aren't any humans alive today who haven't been exposed to advanced human tools so our only option is to engage in a thought experiment. In order to rid yourself of the effects of advanced human tools you need to imagine not only that the Earth is devoid of these tools, you also need to imagine that you are a human who is living in the distant past. You need to imagine that you are living at a time before humans developed the advanced tools of the hunter-gatherers. Can you imagine what it would be like to be a human living at such a time? When I try to imagine what it was like for these early humans the last thing that I imagine is that these humans had a sense that they were special. Rather, I imagine that these humans were regularly killed by non-human animals; that there was a precarious battle for survival; that they lived in a state of fear, of vulnerability. Compared to the killing ability of many of the species of animals which existed in this period the human species was physically inferior – humans were relatively small, soft-skinned and small-toothed; numerous other species had a physical constitution much more suitable for killing. Without advanced tools to defend themselves it seems more likely that these humans might have had *a sense that their predators were the life-forms which were special.* However, it is perhaps more likely that these early humans had neither a sense that their species was special nor a sense that any other species was special. If there are no advanced human tools then a sense of human specialness doesn't arise. So, I'll repeat what I said a little earlier. *The source of the widespread sense of human specialness which exists today is advanced tool use.*

7.2 Do humans have a unique attribute?

There is an important difference between *the source* of the widespread sense of human specialness and the *explanations* that individual humans come up with to justify their sense that the human species is special. The 'sense of specialness' which I have been describing is not something which arises from a process of rationalisation/deliberation/thought – it is, I have claimed, something which almost automatically arises within

humans when they attain a certain level of knowledge concerning their surroundings.

If one asks a large number of humans why they believe that
the human species is special one can expect a multitude of
answers: humans are the only life-forms that have souls;
humans are the only life-forms that have language; humans are
the only life-forms that can think; humans are the only life-
forms that can feel pain/have emotions; humans are the only
moral life-forms; humans are the only life-forms that have
awareness; humans are the only tool-users; humans are the only
life-forms with culture. In their attempt to justify their sense of
human specialness it is amazing what a human can come up
with! Whilst the *sense of* human specialness arises from
advanced tool use, most humans *rationalise* that humans are
special because they have one or more unique attributes.

The widespread belief that humans are special because they
possess a unique attribute is situated within the 'static view';
humans are comparing themselves with other animals in the
present moment.[84] In this section my aim is to convince you that
none of the rationalisations that humans typically make about
why the human species is special are true.

We concluded in *Chapter One* that humans have unique
attributes but that there are a plethora of other species which
have unique attributes too. For example, some whales are
uniquely able to communicate halfway around the world via
their songs. Hummingbirds are uniquely able to hover in the air

---

[84] In other words, the widespread belief that humans are special because they
have a unique attribute has nothing to do with the evolutionary perspective. The
belief is simply: *compare a human with a non-human – the human is special
because they have a unique attribute (A*).*

by being able to flap their wings 80 times per second. And sharks are uniquely able to detect electrical pulses in the ocean up to a mile away because of their special brain cells.

Our concern is whether humans are special – whether there is a fundamental division between humans and the rest of the planetary life-forms. For a unique human attribute to give rise to such a division it would have to be a really special attribute. Such an attribute would need to warrant talk of a *great chasm with humans on the one side and the rest of the planetary life-forms on the other*. If humans are the unique possessors of such a 'special' attribute then the human species would be special. In order for such a 'chasm' to exist humans need to be the only planetary life-form to have the 'special' attribute. If humans have an attribute to an advanced degree and non-humans have the same attribute to a less advanced degree then there is no chasm; there are simply different degrees of development of a jointly-shared attribute.

One of the most obvious candidates for such a 'special' unique human attribute is the soul. One could believe that humans are special because all humans have a soul *and* no non-human planetary life-forms have a soul. If this were true then clearly the human species would have a very 'special' unique attribute; the soul would be an attribute which causes there to be a great chasm; on one side of the rift are the ensouled humans, on the other side of the rift is all of the other planetary life-forms.

Let us accept that every human has a soul. I take this possibility to entail that there are two parts to every human – a biological part and another part which still exists when the biological part ceases to function.[85] If every human has a soul does it follow that there is a great chasm? Of course it doesn't. If one adopts the evolutionary perspective and accepts that the human species evolved from a different species, then one has a serious issue to consider. If humans have souls then does the species which humans evolved from also have souls? If one accepts that the human species evolved out of a different species, and also believes that every human has a soul, but denies that the precursor species had souls, then one has some explaining to do. One needs to assert that at one moment all of the species on the planet existed without any souls, and the next moment, when the first human evolved, that they were somehow accompanied by a soul. From a rational perspective this is hard to accept. I am happy to accept the possibility that humans have souls on the condition that at least one non-human species also has souls. This seems to be the most reasonable position one can take on the matter. If this is right then the possession of a soul is not a unique human attribute. Either humans are not the only species to have souls, or there is no such thing as a soul.

Which other attributes could one believe to be uniquely human and 'special'? In the recent past many humans believed that the ability to use tools was a unique human attribute. However, there is now a plethora of evidence that numerous non-human species use tools: primates, birds, elephants, dolphins, otters and octopuses. It is obviously true that tool-use is not a unique human attribute. Of course, it is also obviously true that humans can use tools in a much more advanced way

---

[85] If you have a different conceptualisation of 'soul' the following considerations will still apply.

than any other species; there are various degrees of tool-using advancement but there is no chasm.

What about the ability to feel pain and the having of emotions? We explored these phenomena at length in *Chapter Five*. Anyone who has spent any length of time observing cats, dogs, chimpanzees, and numerous other species, will surely conclude that it is blindingly obvious that these creatures feel pain and have emotions. This conclusion is given rational backing from the evolutionary perspective; I find it very hard to accept the idea that the first human could feel pain and have emotions but that its non-human ancestor was wholly painless and emotionless. Let us accept that 'feeling states' and emotions are not unique human attributes.

What about the ability to think, to rationalise? We considered this phenomenon in *Section 4.2.8*. It is surely the case that dolphins, chimpanzees, bonobos, and a plethora of other species think. These non-human animals have been observed solving novel problems and acting in ways that can only reasonably be explained if they are thinking. Indeed, one reasonable conclusion to reach would be that every animal that has a brain thinks because *a brain is a thinking thing*. Our present concern is only to note the likelihood, bordering on certainty, that there is one animal that is not human that thinks. After analysing the brain structure of dolphins some humans have even claimed that dolphins have a more advanced ability to think than humans. Thinking is clearly not a unique human attribute.

How about the attribute of awareness? We considered this phenomenon in both *Chapter Four* and *Chapter Six*. It might seem obvious to you that non-human animals such as cats, dogs, dolphins, chimpanzees, and a plethora of other species, have the attribute of awareness. And if it does you are probably correct;

indeed, in *Chapter Six* I defended the idea that awareness exists wherever 40-Hz oscillations exist, and this entails that awareness exists in a plethora of species. It only requires a single non-human life-form on the Earth to have the attribute of awareness to mean that awareness is not a unique human attribute; there is surely at least one such life-form.

What about self-awareness? Recall that I have defended the view that awareness is level-less and has variable 'contents', so to talk of self-awareness is simply to talk of a thought and a state of awareness. In *Section 6.8.1.2* we saw that human self-awareness is a state in which a human *becomes aware of themselves as a distinct entity apart from all the rest of the universe.* So, given that non-human animals contain states of awareness and states of thought, the question is whether or not there is a non-human animal that has been aware of the thought: *I am a distinct entity apart from all the rest of the universe.* Given the high levels of reasoning that have widely been observed in numerous species of non-human animals, and the way that certain non-human animals such as chimpanzees respond to the 'mirror test' (the experimenter puts a red spot on the forehead of a chimpanzee and then shows them a mirror, the chimpanzee responds by touching its own head and not the mirror), the most reasonable conclusion to reach is that some non-human animals possess the attribute of self-awareness.

What about the ability to use language? In order to justify their sense of human specialness it is common to hear humans assert that humans have *language* but that all non-human species only have the *ability to communicate.* This assertion is, no doubt, heavily influenced by the fact that humans have nimble fingers to write with, printing presses to produce books and newspapers with, and a plethora of libraries. If divers were to find a dolphin library then one would probably conclude that there was a dolphin language! Of course, dolphins don't have fingers with which to write or underwater printing presses, so such a library won't be found. But, the lack of such things obviously doesn't mean that dolphins don't have a language.

It is undeniable that many species of non-human animals communicate in very complex ways. If humans were able to fully understand these communication systems then it would perhaps be obvious that some of these species have *their own languages*. When one thinks about it, it is pretty amazing that there are some non-human animals that are able to communicate with humans in the *human language*. Two of the most famous such non-human animals are Alex and Koko. Alex was an African Grey parrot who was able to speak and manipulate the English language. Koko is a gorilla who lives in California; he has mastered 1000 words in American Sign Language and is able to combine them in novel ways. Furthermore, it should be kept in mind that chimpanzees have been observed using symbolism.

Humans do seem to have an advanced language, and to have the unique ability to store this language in a diverse range of ways so that it can be accessed by future generations (e.g. books, blogs, and DVDs). However, I assume that one will agree with me that there is no great chasm with the language-using human species on the one side, and the rest of the planetary life-forms on the other. Humans, perhaps, simply have a language with a complex grammar.

How about morality? Humans often rationalise that they are special because they are 'moral'; however, there is now ample observational evidence that some species of non-human animals have this attribute too, they have a sense of fairness, a sense of right and wrong. Primatologists, such as Professor Frans de Waal, have concluded that acts of consolation and empathetic behaviour are universal amongst the Great Apes. The existence of morality in primate societies should not surprise one. After all, group living entails shared values and requires individuals to take into account the needs of other members within the group. It is likely that morality exists in a number of species of non-human animals. However, the existence of the attribute in a single non-human species means that it is not a unique human attribute.

Finally, let us consider culture. What exactly is culture and is it something that is uniquely human? Culture is a term that humans typically use to refer to some of their activities. Indeed, it is a term that has traditionally been used to create a separation between humans and non-human animals: *culture is what humans have and non-humans don't*. This obviously won't do. The actual activities of humans that are 'cultural' activities need to be specified so that other species can be closely monitored to see if they also partake in these activities. Such observations have revealed that symbolism, tool use, social conformity and learned behaviour exist in chimpanzee societies. So much evidence has been amassed that now numerous primatologists and many anthropologists have concluded that chimpanzees have culture. As the monitoring of other species continues in the future it is likely that that several other species will also be identified as having culture. Culture isn't a unique human attribute.

We have covered a diverse range of possible 'special' unique attributes which could make humans special, and we have concluded that no such special attribute exists.[86] In other words,

---

[86] A couple of days after I finished writing this chapter I bumped into a friend who was seemingly in a state of near-exasperation. He was raising his arms up to his head and as he did so he said in a desperate voice: "What is a self? What is a

there is no great chasm between humans and the rest of the Earth. It is possible that you might disagree with this conclusion.[87] You might believe that there is an attribute which makes the human species special that I have omitted, or that one of the attributes that I have considered is uniquely human. Let us assume that one of the attributes that I have considered *is* actually uniquely human. There is a fundamental issue which we haven't addressed yet. I have considered the attributes that *humans typically rationalise* as being unique human attributes which make humans special. But, why should these attributes *actually* be special? Why should these attributes be 'special'

---

self? I mean *what* is a self?" As I had just finished writing this chapter I was immediately struck by the realisation that I had been considering all of the main candidates for special unique human attributes but I hadn't once used the word 'self'. A moment later I realised that there were other words such as 'subject' and 'person' which I hadn't used. In the ensuing conversation it emerged that he had been reading a book which was describing selves, subjects, experiences, people, humans and the relation between them. I think I, at least partially, managed to alleviate his sense of exasperation by explaining that humans like to create a plethora of unnecessary words. If we discarded 99 per cent of our words we, and he, would be in much less of a state of confusion. I explained that the word 'human' is useful to refer to a part of the universe that typically has arms, legs and a head, etc.; that it useful to have the word 'awareness' to refer to the fact that parts of the universe are aware; and that it is useful to have the word 'thinking' to refer to the fact that parts of the universe think. I think he then realised that 'self', 'person', 'subject', and similar unhelpful creations, are surplus to requirements.

[87] Indeed, given that you are a human who has evolved surrounded by advanced human tools, you will have a sense of human specialness which you seek to rationalise in different ways; it is therefore likely that you will disagree. However, if you have fully comprehended and reflected on what I have said then you might agree.

ones whilst the unique attributes of non-human animals are not special?

It seems obvious that unless one takes the evolutionary perspective there are no special attributes. Why is this obvious? It is simply because from the 'static view' there is no intellectually defensible way of attributing 'specialness' to particular attributes.[88] There are a just a multitude of *diverse* attributes. A human *might believe* that their language is a 'special' attribute; a hummingbird *might believe* that their ability to hover in the air by being able to flap their wings 80 times per second is a 'special' attribute; and so on. These believings are just believings – the reality is that from the 'static view' there are no special attributes. This lack of special attributes leads us to the same conclusion that we reached a little earlier: *there is no great chasm between humans and their surroundings.*

However, if one believes that the universe is an evolving entity then one can believe that certain attributes are special due to their position within the evolutionary process. This will be our concern in the following chapter. I will be contending that the human species has a special place within an evolving universe, and that this entails that *there is a great chasm between humans and their surroundings.*

---

[88] Perhaps this is a little strong. One could get 'special' attributes not from the evolutionary perspective, but from the belief that the universe had a creator. One could believe that when the creator created the universe they created it with certain 'special' attributes – attributes which the creator thinks of as 'special'. I wouldn't wish to deny the possibility that the universe had a creator. However, I don't find the idea that the creator created a 'static universe' to be plausible. If the 'special' attributes of the creator evolved – then we are in the realm of the evolutionary perspective; and, our current concern is whether there are special attributes in the absence of the evolutionary perspective.

Chapter 8

Does the human species have a special place in the evolutionary process?

In *Chapter Two* we considered the phenomenon of evolution and concluded that we will forever lack the knowledge that would enable us to fully understand the nature of both biological evolution and non-biological evolution. In reaching this conclusion we contrasted various views of *evolution as mechanism* and *evolution as path*. The question of evolutionary paths arose again at the end of *Chapter Seven;* in this chapter we concluded that in the absence of such a path humans are not 'special'. That the human species might have a special place in the evolutionary path of the universe was hinted at in *Chapter One* where I suggested that the universe might have evolved humans in order to gain specific abilities.

The idea that the universe evolved humans in order to gain specific abilities implies that there is purpose in the universe. By 'purpose' I mean that the universe has a particular aim/objective/goal. And to say that the universe has a particular aim/objective/goal is simply to agree with Aristotle (cited in Skrbina, 2005, p.46) that:

There is something divine, good, and desirable... [that matter] desire[s] and yearn[s] for

The purpose of the universe is to attain a 'good and desirable state', to attain that which is 'yearned for'. So, to say that the universe brought forth humans in order to gain specific abilities is to say that the universe 'yearned for' humans in an attempt to reach a 'good and desirable state'. It was a specific part of the universe that brought forth the human species – non-human life forms. So, one can reasonably believe that life has a purpose, and that life's purpose is part of the universe's purpose.[89] For life to have a purpose is for life to have a particular aim/objective/goal.

I will be contending that the universe is purposeful, and that the human species has a special place within the evolution of the universe. However, the human species could have a special place in the evolutionary process even if the universe is not purposeful. If the human species evolved as some kind of 'fluke', if the universe did not bring forth the human species in order to gain special abilities, then the human species could still have special abilities. So, whilst much of that which follows will be presented in terms of a purposeful universe, if one is not convinced that the universe is purposeful, one can still accept the main conclusion which I seek to convince you of – that the human species has a special place in the evolutionary process.

If the universe is purposeful then its purpose is to reach a particular state; let us refer to this state as U*. There will be an evolutionary trajectory which leads to U*. If the path that non-

---

[89] By 'life' I don't mean individual life-forms, I mean the totality of life on a planet (the Earth has a living part and a non-living part). Recall that in *Section 2.5* I suggested that life is separated from non-life because the 'biological' part of the universe is a state of 'excitation' or 'exhilaration' (now we can see that this is so because what was 'yearned for' was achieved).

biological evolution and biological evolution has taken could have been radically different then it is hard to make any sense of the idea that there is purpose in the universe. So, if there is purpose in the universe this implies that things evolve according to a rough trajectory; a particular movement along this trajectory towards U* will be the fulfilment of a 'mini purpose'.[90] To say that the purpose of the universe is to reach U* is not to say that the universe will actually attain U*. It is common for purposes to not be achieved – for objectives to not be attained.

What are the possible candidates for U*? What could the purpose of the universe be?[91] There is one obvious candidate: life. Life seems to be an interesting state for the universe to be in. The early universe was seemingly wholly devoid of life. Perhaps the purpose of the universe is to attain a state of life. So, U* could simply be: *life*. Wherever life exists in the universe, this can be thought of as a 'good' state of the universe. According to this view the best state for the universe to be in would be pervaded with life. This means that:

$$U^* = \text{life exists throughout the universe}$$

---

[90] The idea that the universe – which includes human culture – evolves according to a rough trajectory has recently been defended by Robert Wright (*Nonzero*), Simon Conway Morris (*Life's Solution: Inevitable humans in a lonely universe*), Edward O. Wilson & Charles J. Lumsden (*Genes, Mind & Culture: The coevolutionary process*), and Pierre Teilhard de Chardin (*The Phenomenon of Man*).

[91] It is worth noting that the question of whether the universe is purposeful is unrelated to the question of whether the universe had a creator. A purposeful universe is compatible with no creator, a creator which no longer exists, pantheism, panenthesim and theism.

All of the movements of the universe which contribute to U* will be 'mini purposes'; they will be movements which in themselves are purposeful because they contribute to the fulfilment of U*. In the rest of this chapter I will assume that the purpose of the universe – U* – is to bring forth and maintain a state of life in as much of the universe as is possible.

The question of whether the human species has a special place in the universe is one of the most fundamental aspects of the relationship between humans and their surroundings. Are humans an important part of a much larger evolutionary progression which has been ongoing for millions of years? If the human species has a special place in such a progression – if it has a special role in achieving U* – then we have grounds for believing that the 'sense of specialness' which we explored in *Section 7.1* is justified. In other words, we have grounds for believing that there is a chasm between humans and their surroundings. If there is no such special place then the 'sense of specialness' would be misplaced and there would be no 'chasm'.

In this chapter I am proposing that the human species does have an extremely special place in the evolutionary process; I will be claiming that the human species is actually the pinnacle of the evolutionary progression of life on Earth. This is not a popular view because it is widely taken to be 'unscientific' to believe that the evolution of life has a preordained trajectory. However, we have already seen in *Chapter Two* and in *Chapter Three* that there is actually nothing 'unscientific' about such a view; for, science has its limits and we are now on territory which can neither be 'scientific' nor 'unscientific' because it is territory which is forever outside the realm of science. There is no scientific way of knowing whether the entire universe has interior states which propel its evolution along a particular trajectory; science cannot refute Aristotle's belief that the universe yearns and desires to reach a 'good' state.[92]

If the human species is the pinnacle of the evolutionary progression of life on Earth then this should be extremely easy to spot; I am not suggesting that there are any 'undercover operations' – the chasm between the human species and the rest of the planetary life-forms should be so great that it is there for all to see – literally staring anyone who cares to look right in the face. It is this 'staring in the face' which gives rise to the 'sense of human specialness' which I outlined in *Section 7.1*. Humans are technological animals. Whether one is a human, or an alien visitor to the Earth, one cannot evade this truth – there is no hiding place. If you are human then being surrounded with advanced tools and technology gives rise to a 'sense of human specialness'. If you are an alien visitor who observes the Earth from your UFO what are you going to report to your fellow aliens when you return to your home planet? You will report: *there are lots of life-forms on the Earth and one of them is different – one of them is technological.* Of course, being technological is not itself a 'special-making' attribute. Creating technology is only a 'special-making' attribute if technology has a special place in the evolutionary process. If humans are special, if they are the pinnacle of the evolution of life on Earth, it is because they are the bringers forth of technology.

---

[92] And, as will soon become clear, my claim is not that there is a preordained trajectory towards 'biological humans' (such a claim would be much more contentious); my claim is simply that there is a preordained trajectory towards 'technology creators' (technology is the zenith of advanced tool use). Recall that there are two senses of 'human'. The trajectory is not towards a particular biological form ('biological humans') it is towards 'technology creators'=doubters of their naturalness ('humans').

At this point one needs to take a pause and reflect. I have contended that life on Earth has 'yearned for' the human species. Yet, the dominant contemporary view is that the rest of life on Earth would be far better off if the human species were to become extinct! As the human species has spread out over the entire planet one effect has been that many species have become extinct, and unfortunately many more are in grave danger of extinction. The human species has turned natural habitats into 'concrete jungles'; it has initiated mass deforestation and mass agriculture; through its web of trade and transport links it has imported alien species into inappropriate habitats which have decimated the native species; and it has released large amounts of oil into the oceans with disastrous effects. Now, many see the threat of human-induced global warming as the final 'nail in the coffin' of other species.

According to this view, the human species through its selfish desire to plunder the world's resources in order to have a high standard of living is set to destroy a multitude of species. The human species is the 'destroyer of life' and if it were to become extinct this would be a great event for the rest of life on Earth. The advocates of this view believe that a couple of millennia after the extinction of the human species the biological diversity of the Earth would be vastly higher; life would supposedly be flourishing in the absence of the destructive humans. This view is grounded in what has happened in the past. In the past when there have been mass extinctions of life on Earth it *has* been the case that after a long enough period of time biological diversity has become just as rich as it was before the mass extinction. However, one cannot always use the past as a guide to the future.

This dominant view is not totally wrong but from the evolutionary perspective one can see that it misses out on the bigger picture. In the bigger evolutionary picture the human species is not 'the destroyer'; to the contrary, the human species is 'the saviour of life'. If the 'saviour' were to become extinct this would clearly *not* be in the interests of life on Earth. The human species is the saviour because it is the bringer forth of technology which life/the universe has yearned for in its attempt to reach U*.

There is a short and a long story to tell concerning the cosmic significance of human technology (and therefore of the 'specialness' of the human species). Here is the short story which has five premises. Firstly, life is a good state for the universe to be in. Secondly, life yearns to survive (it strives to survive). Thirdly, the universe and the Earth are evolving and in the future the Earth will not be able to sustain life; it is well known that the Sun, which enables life to exist on the Earth, will also become the enemy of life on Earth, it will get increasingly hotter until it explodes. Fourthly, life obviously needs to evolve a technological species if it is to survive.[93] Fifthly, the human species, as that part of life on Earth which has become technological, is the saviour of life on Earth. So, the conclusion is that

---

[93] It is possible that you might not be overly comfortable with the phrase 'life needs'. The premise is obviously true because the Earth will not be able to sustain life in the future, but how does life evolve in such a way that its needs get fulfilled? I believe that these needs get fulfilled because there are motivations and 'feeling states' within life-forms which propel evolution along a rough trajectory towards a technological species (this process is grounded in the Aristotelian 'yearning'). As we saw in *Section 2.2* according to the symbiogenetic evolutionary mechanism biological evolution is propelled by "reasons" through "prototaxis". These "reasons" are grounded in 'feeling states'.

the human species has a special place in the evolutionary process.

Let us move to the longer, and more contentious, story.[94] The short story apparently only applies in the distant future. However, the long story involves the present and the immediate future; it entails that the environmental crisis and human-induced global warming are actually positive events in the evolution of life on Earth. The heart of this more complex story is the realisation from Earth Systems Science that since life arose on the Earth it has been homeostatically regulating the conditions of the Earth to keep them favourable for complex life. The key way to appreciate this is by realising that the amount of solar energy reaching the Earth from the Sun has increased by 40 per cent since life arose on Earth. On hearing this one would probably expect that the temperature of the Earth's atmosphere has increased over this period. But it hasn't because life, in tandem with the rest of the Earth, has been regulating the temperature of the atmosphere in order to keep it favourable for its continued existence. As Sir James Lovelock explains:

> right from the beginning of life, around three and a half aeons ago, the Earth's mean surface temperature has never varied by more than a few degrees from its current levels. It has never been too hot or cold for life to survive on our planet, in spite of drastic changes in the composition of the

---

[94] I tell this story in detail in my book: *"Is the Human Species Special?: Why human-induced global warming could be in the interests of life"* (Cranmore Publications, 2010). It is further developed in my book: *"What Does it Mean to be Green?"* (Vitae Publications, 2011).

early atmosphere and variations in the sun's output of energy.

(Lovelock, 2000, p.48)

If our planetary temperature depended only on the abiological constraints set by the sun's output and the heat balance of the Earth's atmosphere and surface, then... all life would have been eliminated.

(Lovelock, 2000, p.20)

[Gaia] is old and has not very long to live. As the sun grows ever hotter it will, in Gaia's terms, soon become too hot for animals and plants and many of the microbial forms of life.

(Lovelock, 2006, p.46)

only for a brief period in the Earth's history was the sun's warmth ideal for life, and that was about two billion years ago. Before this it was too cold for comfort and afterwards it has progressively grown too hot... The sun is already too hot for comfort.

(Lovelock, 2006, pp.44-5)

The ability of life/Gaia to offset the increasing output of the sun and keep the Earth favourable for life is weakening because the "sun is already too hot for comfort". Lovelock (2006, p.45) claims that: "The brief interglacials, like now, are, I think, examples of temporary failures of ice-age regulation." In short, the non-technological ability of the Earth to sustain life is weakening and these "temporary failures" will be followed by a complete failure/elimination of life, unless life brings forth the technology which can sustain life by regulating the atmosphere. This technology is required to keep the temperature of the atmosphere down in the face of the increasing output of the ageing Sun. In other words, if life is to survive there needs to be a transition from non-technological regulation of the atmosphere to technological regulation of the atmosphere.

The need for such a transition to technological regulation is an imminent one because of the perturbations that have been made to the homeostatic regulatory capacity by the human species.[95] Before humans evolved, carbon from the atmosphere was stored under the land surfaces of the Earth in order to keep the temperature of the atmosphere down in the face of increasing solar output. Humans have released the majority of these stores of carbon but, so far, the effects of this have barely manifested themselves. As Stephen Peake (2003, p.78) explains, "Even if we engage in further actions to reduce GHG emissions, the [IPCC] models tell us that anthropogenic climate change will continue for centuries." One of the main reasons for this is that colossal amounts of carbon that humans have released have become temporarily stored in the deep ocean thermohaline

---

[95] On the purposeful view of the universe these perturbations were 'inevitable' – they are required for U*. These perturbations are a side-effect of the development of technology and they ultimately result in humans technologically regulating the atmosphere of the Earth for the benefit (the survival) of life on Earth.

circulation, and it will eventually resurface.[96] As Robert H. Stewart (2009, p. 214, p.233) explains:

> New CO_2 is released into the atmosphere when fossil fuels and trees are burned. Roughly half of the CO_2 released into the atmosphere quickly dissolves in the cold waters of the ocean which carry it into the abyss...therefore temporarily reducing atmospheric CO2. Eventually, however, most of the CO2 must be released back

After the carbon has sunk into the thermohaline circulation it takes from 100 years to 1000 years to re-enter the atmosphere. This means that almost all of the carbon that has become stored since the start of the industrial revolution has yet to re-enter the atmosphere. When this carbon starts to be released *en masse* then the temperature of the Earth's atmosphere is likely to shoot upwards. The initial increase in atmospheric temperature could easily trigger other large-scale discontinuities such as a runaway greenhouse effect resulting from the destabilisation of methane clathrate reservoirs. By the year 3000 – when all of the carbon that is currently stored in the thermohaline has been released – the conditions of the Earth could easily not be suitable for human habitation (this applies even if humans were to stop using fossil fuels today; our continuing intoxication with fossil fuels is just bringing forth the 'inhabitability date'). This is not inevitable. It is possible that humans might have fulfilled their purpose before this happens; humans might be technologically regulating the temperature of

---

[96] Dense cold water sinks at high latitudes and travels through the ocean until it eventually reaches the northern Indian Ocean and the northern Pacific Ocean.

the atmosphere of the Earth, for the benefit of life on Earth. If so, then when the carbon emerges from the thermohaline humans can maintain the temperature of the atmosphere. One should reflect on this:

> The continued existence of the human species in the near future depends on developing the technological ability to regulate the temperature of the atmosphere of the Earth.[97]

This 'long' story can be summarised as follows. Firstly, the Earth and the Sun, like all other parts of the universe, are ageing entities. Secondly, the universe is divided into two parts – life and non-life. Thirdly, life is a good state for the universe to be in. Fourthly, life, and complex life in particular, require certain conditions in order to survive. Fifthly, when life arises it strives to stay in existence by spreading out over the planet it arises on and by regulating the temperature of that planet's atmosphere to keep it favourable for its continued existence.[98] Sixthly, as the Sun's energy keeps on increasing the point will come when, in the absence of a technological species, the ability of planetary life to regulate the temperature of the planet's atmosphere will

---

[97] The other option for saving the human species (and non-human life-forms) – that humans use technology to move to another planet – seems much more difficult to attain in the available time. However, this option also seems likely to be fulfilled within a few thousand years (presuming that we successfully regulate the temperature of the Earth's atmosphere in the meantime).

[98] Life has done this on the Earth. Some humans find this hard to comprehend and/or accept because they assume that regulating the conditions of the Earth requires an intention to so regulate. Regulation occurs without an intention to regulate; this is because the striving for survival itself leads to the outcome that is the regulation of the conditions of the Earth.

cease.[99] Seventhly, in order to survive life needs to evolve a technological species. Eighthly, developing technology involves a weakening of the homeostatic regulatory capacity (the release of fossil fuels, human-induced global warming). Ninthly, on the Earth the human species is that part of planetary life which is technological. Tenthly, the purpose of the human species is to be the saviour of planetary life through developing and deploying the technology which regulates the temperature of the Earth's atmosphere. Finally, human-induced global warming is the catalyst which causes the human species to fulfil its purpose. So, the human species is special.

One might find all of this to be slightly unconvincing, even far-fetched.[100] So, it might be a good time to reveal that I have found an interesting ally in the form of the German romantic Friedrich Hölderlin. Hölderlin died before the human species realised that they had severely perturbed the environment of the Earth. However, his philosophy, when viewed from the perspective of the 'environmental crisis' of modernity, provides a very interesting perspective on the relationship between the human species and their surroundings in an evolving universe. It is a view according to which humans and their technology have an important role to play in the evolutionary progression of life on Earth.

---

[99] This is the basic premise of 'planetary astrobiology' – see: *The Life and Death of Planet Earth: How science can predict the ultimate fate of our world* by Peter Ward and Donald Brownlee.

[100] One should take it seriously though. That is, if one cares about the future survival of the human species, and the continuation of the life that has evolved on the Earth.

On my view, and on Hölderlin's view, the universe can be viewed as a continuous process of unfolding wherein disharmonies bring forth harmonies, which in turn bring forth further disharmonies. The overall trajectory is to a higher state of harmony – a better state of the universe. From another perspective (which I believe Hölderlin would appreciate) one could think of the universe as a kind of game – life arises, and then life tries to spread out as far as is possible; initially life is limited to its host planet, but a technological species can enable it to spread out further. Life's objective in this game is to spread out over the entire universe. Will the game be won? Will it be lost? There are likely to be multiple aspects to the game (life could exist on other planets). If the game is lost in this running of the universe (the events following the Big Bang), will life make more progress next time (if there is a Big Crunch followed by another Big Bang)? My synthesis of Hölderlin's philosophy with the 'environmental crisis' of modernity appears in the following section.

8.1   Human nature, cosmic evolution and modernity in Hölderlin[101]

The German Romantic Friedrich Hölderlin developed a unique perspective on the relationship between humankind and the rest of nature. He believed that humanity has a positive role to play in cosmic evolution, and that modernity is the crucial stage in fulfilling this role. In this section I will be arguing for a reinterpretation of his ideas regarding the position of humankind in

---

[101] This section was published as a paper in the journal *Cosmos and History: The Journal of Natural and Social Philosophy* in 2007 (Vol. 3, No. 1).

cosmic evolution, and for an application of these ideas to the 'environmental crisis' of modernity. This reinterpretation is significant because it entails an inversion of the conventional notion of causality in the 'environmental crisis'; instead of humans 'harming' nature, in the reinterpretation it is nature that causes human suffering.

Hölderlin's ideas are of particular interest because he yearned for an end to human suffering, but was also firmly convinced that humankind was inevitably destined to be separated from nature, and thereby destined to endure suffering. Hölderlin envisioned a positive role for humanity in cosmic evolution, a role which has significant implications for both human nature and cultural evolution. In this section I will be outlining Hölderlin's ideas, and arguing for an application of them to the 'environmental crisis' of modernity. Hölderlin's conception of the human-nature relationship as part of an unfolding process of cosmological change seems to be of great relevance today, an age that is characterized by belief in the meaninglessness of human existence, and by concern about the way that we have altered the pre-human conditions of the Earth. Hölderlin's views provide a unique perspective on modernity that is worthy of serious consideration.

I start by outlining Hölderlin's views on the role of humankind in universal evolution. I then review the secondary literature on Hölderlin that relates to these ideas. I proceed to argue that Hölderlin's philosophy is applicable to, and gives a unique perspective on, the 'environmental crisis' of modernity. I argue that the existing secondary literature on Hölderlin has not recognized this, and that a reinterpretation of the role of humanity in Hölderlin's philosophy of cosmic evolution is therefore required. My central claim is that for Hölderlin, modernity and the related notion of the contemporary 'environmental crisis' is a necessary stage of cosmic evolution, and thus that it is far from a 'crisis'. Rather it is a necessary stage of disharmony that will inevitably be followed by a re-conquered harmony. I will argue that for Hölderlin this disharmony is characterized by the environmental changes that are resultant from the development of technology.

8.1.1 *Hölderlin's philosophy of human nature, cosmic evolution and modernity*

The starting point of Hölderlin's philosophy is that there must be a basic unknowable reality which precedes self-consciousness wherein subjects and objects are not in existence but are both part of a 'blessed unity of being'. Hölderlin (2003, p.191) describes this unity as, "Where subject and object simply are, and not just partially, united...only there and nowhere else can there be talk of being." He argues that the 'blessed unity of being' (which he also refers to as 'nature') is responsible for the coming into existence of humanity through using its power to initiate a division of itself into subjects and objects. This division of being causes the emergence of judgement. Hölderlin (2003, p.192) states that, "'I am I' is the most fitting example of this concept of judgement...[as] it sets itself in opposition to the *not-I*, not in opposition to *itself.*"

The division means that human beings are not capable of actions that are independent of nature; Hölderlin (cited in Stone, 2003, p.423) states that, "all the streams of human activity have their source in nature." It is revealing to compare this claim with the words of Hölderlin's (1990, p.35) character Hyperion, "What is man? – so I might begin; how does it happen that the world contains such a thing, which ferments like a chaos or moulders like a rotten tree, and never grows to ripeness? How can Nature tolerate this sour grape among her

307

sweet clusters?" For Hölderlin, man is the 'violent' being, whose coming into existence in opposition to the rest of nature was *initiated* by nature.

Hölderlin sees this opposition between man and the rest of nature as culminating in modernity – an era that he claims is characterised by the absence of the gods. In *Brot und Wein* Hölderlin (1990, p.185) writes, "Though the gods are living, Over our heads they live, up in a different world...Little they seem to care whether we live or do not." A key question for Hölderlin is how we deal with this separation. He envisions two possibilities – the 'Greek' response which is to dissolve the self and die, and the 'Hesperian' response of a living death.

Hölderlin came to view the 'Greek' response as hubristic, it being based on an anthropocentric desire to oppose the division initiated by nature. He thus sees the 'Hesperian' response of living and carrying out actions that are dependent on nature for their origination as the appropriate non-hubristic response to our separation. Hölderlin's position is that as nature created the separation, *only* nature can bring the separation to an end. He sees this process of separation and reconnection as part of a broader cosmic picture wherein nature is an unfolding organism rather than a huge mechanism. This organismic view enables him to envision teleological processes in nature which give rise to his claim that there will be, "eternal progress of nature towards perfection" (Peacock, 1938, p.36).

8.1.2 *Interpretations of Hölderlin and his concept of fate*

In this section I set out my view of Hölderlin's conception of fate – that all human actions are part of the evolution of nature towards perfection. I do this by reviewing the existing scholarly literature on Hölderlin and showing that whilst these interpretations all recognise parts of Hölderlin's conception of fate that they do not capture the whole of it. I start with interpretations of human nature, move on to cosmic processes, and finally consider the role of modernity within these processes.

At the level of the human there is a general consensus in the literature that Hölderlin's position is that humans are endowed by nature with qualities that shape human nature, and that this inevitably shapes human interactions with the rest of nature. There are various names in the literature for the qualities which are endowed to humans. Dennis J. Schmidt refers to the qualities present in humans as their 'formative drive.' Schmidt (2001, p.139) claims that: "Hölderlin suggests that human nature and practices are to be understood by reference to a formative drive which expresses itself as a constant need for 'art'." In a similar vein, Thomas Pfau argues for an 'intellectual intuition' (knowledge of the 'surroundings in-themselves' that is not attained by one being affected by them). Pfau (1988, p.15) states that: "Hölderlin recasts the convergence of "freedom and necessity" as the most primordial synthesis of intellect and intuition itself, a synthesis which takes place within the subject itself. He thus approaches what Kant had repeatedly ruled out as an "intellectual intuition"."

In agreement with Schmidt and Pfau, Franz Gabriel Nauen (2001, p.139) argues that for Hölderlin, "all men do in fact have the same basic character...all human activity can be derived from the same *elemental drive* in human nature." The 'formative drive'/'intellectual intuition'/'elemental drive' identified in the literature explains why man can be seen as the 'violent' being. Human nature is to engage in 'art', to utilize the resources of nature so that culture can be generated and sustained. This generation of human culture actually benefits nature as a whole, but it requires large-scale modification of parts of non-human nature. The destiny of man is thus a disruptive one. It is clear that it is also an undesirable one. Nauen (2001, p.139) states that for Hölderlin: "Even war and economic enterprise serve to fulfil the destiny of man, which is to "multiply, propel, distinguish and mix together the life of Nature"."

So Hölderlin sees human nature, economic production and even war as parts of a broader cosmic evolutionary process; the universe *as a whole* is seen as evolving to perfection. There will inevitably be aspects of this evolution that from a narrow perspective could be viewed as 'less than perfect'. These negative aspects of the evolutionary process – from war, to the presence of evil in its entirety – have to be seen as inescapable parts of the whole process.

The key point is that for Hölderlin the cosmic evolutionary process *ends* in perfection. Thus, Ronald Peacock (1938, p.22) argues that, "the division produced by conflict is followed by a re-conquered harmony." Similarly, Anselm Haverkamp (1996, p.48) argues that an interpretation of the poems *Andenken* and *Mnemosyne* is the expression, 'where danger threatens, salvation also grows.' Whilst, Martin Heidegger (2003, p.261) translates the opening lines of *Patmos* as, "But where danger is, grows the saving power also." Hölderlin's view is clearly that from a narrow and short-term perspective danger and conflict are often the norm, but that these things actually play a part in bringing about a greater harmony in the future. In the long-term they are all part of the evolution of the whole universe to perfection.

Cosmic evolution is thus one long process of disharmonies and inevitably following harmonies. Peacock (1938, p.22) argues that Hölderlin's vision is of a, "harmonised process of life which comprises within itself the rhythmic movement from chaos to form and back again, and an emotional experience of this which in the sphere of nature knows only the one rapture, but in the human sphere suffering and joy." It is revealing that this interpretation sees 'violent' humans as suffering, whilst nature is purely rapturous. This clearly sheds light on the question posed by Hölderlin's (1990, p.35) character Hyperion: "How can Nature tolerate this sour grape among her sweet clusters?" The answer seems to be that human 'violence' *enables* nature to be rapturous. As part of this rapture humans experience suffering.

Why should suffering be a uniquely human experience? To explain this Peacock (1938, p.36) cites part of a letter from Hölderlin to his brother, "Why can they [humans] not live contented like the beasts of the field? he asks: and replies that this would be as unnatural in man, as in animals the tricks, or arts, man trains them to perform. Thus he establishes that the arts of man are natural to man. Culture, then, derives from nature; and the impulse to it is the characteristic which distinguishes man from the rest of creation."

The human impulse to culture has culminated in the era of modernity. Hölderlin sees this period as one of great significance as he sees it as a historical epoch that is characterised by the *absence of the gods*. To be consistent with his views on harmonised evolution to perfection there must be a reason for this absence. Indeed, Peacock (1938, p.92) argues that Hölderlin thinks that, "a godless age is part of a divine mystery, it is as necessary as day, ordained by a higher power." Furthermore, Heidegger (1956, p.190) claims that the gods are still present, despite their absence: "man who, even with his most exulted thought could hardly penetrate to their Being, even though, with the same grandeur as at all time, they were somehow there."

The absence of the gods in modernity is deeply related to the contemporary danger that exists in modernity. It should be remembered that this danger cannot be a cause for concern for Hölderlin – as all dangers are inevitably followed by regained harmonies. Nevertheless, Heidegger attempts to identify the exact danger that Hölderlin believed is present in modernity. Heidegger (2003, p.261) claims that, "the essence of technology, enframing, is the extreme danger." It must follow that for Heidegger (2003, p.261), "precisely the essence of technology must harbor in itself the growth of the saving power." Heidegger sees this as occurring when the essential unfolding of technology gives rise to the possibility of opening up a "free relation" with technology which is inclusive of non-instrumental possibilities (Scharff and Dusek, 2003, p.248).

In an interpretation of the 1802 hymn *Friedensfeier,* Richard Unger draws out Hölderlin's views on the absence of the gods in modernity (Unger, 1984, pp.100-5). In *Friedensfeier* the entire span of Western civilization is characterised as a thunderstorm which is ruled by a "law of destiny" which ensures that a certain amount of "work" is achieved. Unger (1984, p.102) argues that it is clear that this "work", "is the product of the storm itself and that it designates the harmonious totality of earthly existence during the coming era." The end of the "storm" of modernity enables the arrival of a mysterious "prince" who makes it possible that, "men can now for the first time hear the "work" that has been long in preparation "from morning until evening"." (Unger, 1984, p.101)

Following the inevitable successful accomplishment of the "work" of Western civilization, the great Spirit will disclose a Time-Image which will, "be a comprehensive depiction of the historical process and its triumphant result." (Unger, 1984, p.104). Unger (1984, p.105) argues that, "the Image shows that there is an alliance between the Spirit of history and the elemental divine presences of nature – for the natural elements with which man has always worked have played integral and essential parts in man's history." The triumphant result of the actions of humankind in modernity is clearly an example of a re-conquered harmony that follows division.

In Unger's interpretation of *Friedensfeier* we have a picture of modernity in which humans are carrying out "work" under a "law of destiny". The crucial factor is that humanity is ignorant that it is working under a "law of destiny" in modernity, until modernity has ended. It is then that through the Time-Image the great Spirit reveals the successful outcome of modernity, and the *nature and value* of the accomplished "work". This is a prime example of a short-term and narrow perspective entailing the perception of a lack of destiny and of needless suffering, whilst in the longer-term the same events are seen to be an inevitable part of a broader positive outcome – the evolution of the universe to perfection.

This difference of perspectives can explain an apparent contradiction in the literature between Unger's interpretation of *Friedensfeier,* and Schmidt's analysis of Hölderlin's 1801 letter to Bohlendorff. This letter was written only one year before *Friedensfeier* and Schmidt (2001, p.137) claims that in it Hölderlin's position is, "that the peculiar flow of modernity is the lack of destiny." The apparently contradictory views of Unger and Schmidt can be reconciled through recalling Peacock's (1938, p.92) interpretation that, "a godless age is part of a divine mystery, it is as necessary as day, ordained by a higher power," and comparing it to Unger's claim that men are blind to the point of the "work" that they have been carrying out until the "storm" of Western civilization has passed.

The comparison reveals that the "law of destiny" applies to the activities of *humanity as a collective* in Western history, activities that are ordained by a higher power for a specific purpose. In contrast, the "lack of destiny" applies to *individual human beings*. This difference arises because individual humans are unaware that their actions are part of an inevitably unfolding cosmic plan, it is only the fruition of the plan than enables realization. Instead, humans believe that they have free will and live in a meaningless age. Therefore, modernity can at one and the same time be characterized as both a period governed by a "law of destiny" and a period constituted by a "lack of destiny". The difference is purely one of perspective.

This conception of modernity as simultaneously being a period of a "lack of destiny" and a "law of destiny" raises the issue of anthropocentricism. If human attitudes and actions towards nature are in the interests of nature, then it seems that there is no such thing as a *truly* anthropocentric attitude. The appropriate attitude that humans should take to the objective side of nature, given Hölderlin's philosophy, has been addressed by Alison Stone. Stone (2003, p.424) argues that because, "according to Hölderlin's thinking, we have become separated from nature by *its* power alone, so it is not within *our* power to undo separation." Therefore, according to Stone (2003, p.424): "the appropriately modest response is to endure separation – to wait, patiently, until nature may change its mode of being." Stone (2005) concludes that for Hölderlin a truly non-anthropocentric environmental view of the rest of nature requires: "the *acceptance* of disenchantment, of separation, of meaninglessness."

This view is concordant with the "lack of destiny" perspective. However, when the "law of destiny" is taken into account, then the hidden meaning is revealed. Furthermore, the whole notion of the attitudes of individual humans then becomes irrelevant. It seems that there cannot be such a thing as a *truly* anthropocentric attitude, because all attitudes originate from nature, and they all lead to actions which fulfil the "law of destiny". It may seem that our attitudes to nature are of importance, but this is because we believe in a "lack of destiny", and are inevitably blind to the bigger picture of the "law of destiny". Whatever our attitudes as individuals, our relationship with the rest of nature as a collective would be 'for the best'.

8.1.3 A reinterpretation of the human in cosmic evolution

The interpretations of Hölderlin that I have reviewed all give an accurate representation of Hölderlin's views. However, they are all partial views. They all miss the 'big picture' of what Hölderlin's views imply about what it means to be a human in the context of cosmic evolution, and the consequent implications for the perspective from which we should view modernity and the 'environmental crisis'. In an attempt to fully grasp these implications I am going to defend the thesis that: *Hölderlin's philosophy leads to the conclusion that the 'environmental crisis' is a necessary stage in the purposeful evolution of nature towards perfection.* This is an interesting thesis because, if accepted, it would supplant the conception of the meaninglessness of human existence with a conception of positive cosmic purpose.

The argument I will be making centers on three key aspects of Hölderlin's philosophy. Firstly, that he believes that nature is purposefully evolving towards perfection. Secondly, that he believes that the achievement of this perfection requires human actions. Thirdly, that he believes that human actions are determined by nature. Acceptance of these three claims leads to the conclusion that human actions are determined by nature as a necessary stage in the purposeful evolution of nature towards perfection. As the 'environmental crisis' of modernity is purely resultant from human actions, a second conclusion inevitably follows. This is that the 'environmental crisis' itself is determined by nature as a necessary stage in the purposeful evolution of nature towards perfection.

I will now present evidence to support the three key claims. The first claim is that Hölderlin's belief is that *nature is purposefully evolving towards perfection*. The universe can either be viewed as a giant mechanism or as an unfolding organism; Hölderlin clearly held the latter view. This conception of the universe explains his belief that nature unfolds in a way that serves its own purposes; that disharmonies are followed by regained harmonies. This is why Peacock (1938, p.36, p.105) claims that Hölderlin believed in, "the eternal progress of nature towards perfection," and, "the emergence of perfection in the course of natural development."

This firm belief clashed with Hölderlin's personal yearning for immediate perfection in life. His immense desire to see a morally just world was completely at odds with his philosophical belief that the perfection he sought could only be achieved in the course of natural development. The movement to perfection envisioned by Hölderlin is thus a fatalistic one, an inevitable evolutionary progression towards perfection. Peacock (1938, p.93) captures this with his claim that for Hölderlin there is an "acute sense of 'Fate', of inevitability, expressed again and again in his work. Fate is revealed in the process of history... it is inherent in the passage of form to chaos, and of disintegration to a new harmony."

This first claim is the most straightforward of the three. The second claim is that *Hölderlin believes that the achievement of perfection requires human actions*. The starting point in defending this claim is Hölderlin's central belief that nature *used its power* to divide itself and thereby create humankind. This division means that the split was part of the evolutionary process rather than a random occurrence. We can ask ourselves why this may have been a necessary occurrence. An initial answer seems to be Nauen's (2001, p.139) claim that, "Even war and economic enterprise serve to fulfil the destiny of man, which is to "multiply, propel, distinguish and mix together the life of Nature"."

In *The Perspective from which we Have to look at Antiquity* Hölderlin (1988a, p.39) asserts that: "antiquity appears altogether opposed to our primordeal drive which is bent on forming the unformed, to perfect the primordial-natural so that man, who is born for art, will naturally take to what is raw, uneducated, childlike rather than to a formed material where there has already been pre-formed [what] he wishes to form." In a letter to his brother he also asserts that: "the impulse to art and culture…is really a service that men render nature." (Peacock, 1938, p.37).

The source of Hölderlin's primordeal drive to art is nature, because it is nature that created us and endowed us with our capabilities. This is clear from Peacock's (1938, p.37) interpretation that, "Man cannot be master of nature; his arts, *necessary though they may be in the scheme of things,* cannot produce the substance which they mould and transform; they can only develop the creative force, which in itself is eternal and not their work."

Hölderlin's primordeal drive to art in humans has inevitably led to the epoch of modernity. Human actions in this epoch appear to be central to the achievement of perfection. Hölderlin claims that modernity is an epoch that, "is as necessary as day, ordained by a higher power." (Peacock, 1938, p.92). Furthermore, humans have been involved in "work" in modernity that is clearly constitutive of the importance of the epoch. This is clear from Unger's interpretation of *Friedensfeier* in which the "law of destiny" ensures that a certain amount of human "work" is done.

The crucial factor is that humanity is ignorant that it is working under a "law of destiny" in modernity, until modernity has ended. It is then that through the Time-Image the great Spirit reveals the successful outcome of modernity, and the nature and value of the accomplished "work".

There is no doubt that in Hölderlin's view human actions and their resultant "work" in modernity are part of purposeful evolution to perfection. What is interesting is the exact nature of the "work". There is an obvious connection between the "work" of modernity (*Friedensfeier*) and the "danger" we face in modernity (*Patmos*). Heidegger's (2003, p.261) interpretation of *Patmos* that, "the essence of technology, enframing, is the extreme danger," makes it clear that the "work" of modernity is the development of technology. In fact, technological development in modernity seems to be the culmination of Hölderlin's primordeal drive to art. Furthermore, it is very hard to think of any other distinctive aspects of modernity that are resultant from human actions, present an extreme danger, and have cosmic significance. Therefore, for Hölderlin, the achievement of perfection seems to require the human development of technology.

It is interesting that Heidegger sees the danger we face from the "work" of modernity as the essence of technology rather than actual technology. Andrew Feenberg (2003, pp.327-337) has criticised Heidegger for this abstract concentration on essences rather than the actual technology itself. A "Feenberg interpretation" of *Patmos* seems to be more in accordance with Hölderlin's views than the "Heidegger interpretation", as Hölderlin's philosophy is grounded in actualities rather than essences. Hölderlin sees a positive role for actual technology in cosmic evolution; this means that *actual technology* has a cosmic purpose. Therefore, it seems that both the danger we face, and the saviour, must be the *actual* technology developed by human actions.

The importance of the human split from the rest of nature can also be seen in the words of Hölderlin's (1990, p.123) character *Hyperion:* "How should I escape from the union that binds all things together? We part only to be more intimately one, more divinely at peace with all, with each other. We die that we may live." Human actions are thus depicted as a 'living death' that is necessary for the life (and continued movement to perfection) of nature as a whole. This explains Peacock's (1938, p.22) interpretation that, "the sphere of nature knows only the one rapture, but in the human sphere [there is] suffering and joy."

The third claim is that *Hölderlin believes that human actions are determined by nature.* There are many passages in Hölderlin's novel *Hyperion* that attribute the responsibilities for human actions to a power or god: "There is a god in us who guides destiny as if it were a river of water, and all things are his element." (1990, p.11).....oh forgive me, when I am compelled! I do not choose; I do not reflect. There is a power in me, and I know not if it is myself that drives me to this step." (1990, p.79) .....I once saw a child put out its hand to catch the moonlight; but the light went calmly on its way. So do we stand trying to hold back everchanging Fate. Oh, that it were possible but to watch it as peacefully and meditatively as we do the circling stars." (1990, p.22) .....Man can change nothing and the light of life comes and departs as it will." (1990, p.127) .....We speak of our hearts, of our plans, as if they were ours; yet there is a power outside of us that tosses us here and there as it pleases until it lays us in the grave, and of which we know not where it comes nor where it is bound." (1990, p.29).

Hölderlin's belief in the lack of human free will is perhaps clearest in his claim in a letter to his mother regarding the views of Spinoza that, "one *must* arrive at his ideas if one wants to explain everything." (Holderlin, 1988b, p.120). Spinoza's ideas are summed up by Moira Gatens (1996, p.111) as: "Nature in all its aspects is governed by necessary laws, and human being no less than the rest of nature is determined in all its actions and passions, contrary to those who conceive of it as 'a dominion within a dominion'."

In order to make abundantly clear Spinoza's - and thus Hölderlin's – views on a lack of human free will here are two quotes from Spinoza (1966, pp.294-5): "I say that thing is free which exists and acts solely from the necessity of its own nature...I do not place Freedom in free decision, but in free necessity." And, "a stone receives from an external cause, which impels it, a certain quantity of motion, with which it will afterwards necessarily continue to move...Next, conceive, if you please, that the stone while it continues in motion thinks, and knows that it is striving as much as possible to continue in motion. Surely this stone, inasmuch as it is conscious only of its own effort, and is far from indifferent, will believe that it is completely free, and that it continues in motion for no other reason than because it wants to. And such is the human freedom which all men boast that they possess, and which consists solely in this, that men are conscious of their desire, and ignorant of the causes by which they are determined."

Furthermore, in an interpretation of Hölderlin's *Stutgard*, Peacock (1938, p.25) argues that, "the laws of growth govern the culture as well as the lives of men...the one process comprehends all things and the one rhythm manifests itself again and again...in the progress of history; in the spiritual life of individuals." In this vision not only human nature, but also the evolution of culture, is seen as an inevitable historical progression. Peacock's (1938, p.90) interpretation of Hölderlin is that, "man's spirit is but part of the One Spirit," which Hölderlin insists is involved in a "movement...through successive historical generations" (Peacock, 1938, p.114). The spirit of man is thus governed by the larger Spirit of nature. This is the sense in which, "all the streams of human activity have their source in nature." (Stone, 2003, p.423).

The nature of the relationship between man's spirit and the Spirit of nature is made clear in the following quote from Hölderlin's (1990, p.122) character Diotima: "a *unique destiny* bore you away to solitude of spirit as waters are borne to mountain peaks." This concept of individual humans having a unique destiny was the view of Johann Herder, who was one of Hölderlin's inspirations. Herder saw nature as a great current of sympathy running through all things which manifested itself in unique inner impulses within different individuals. This means that every human has a unique calling – an original path which they ought to tread. According to Herder (cited in Taylor, 1994, p.375), "Each human being has his own measure, as it were an accord peculiar to him of all his feelings to each other." Clearly, for both Herder and Hölderlin, human actions at any one time are determined in accordance with the movements of the One Spirit of nature.

I have presented evidence for the claims that for Hölderlin: *nature is purposefully evolving towards perfection, the achievement of this perfection requires human actions, and human actions are determined by nature.* Acceptance of these three claims leads to the conclusion that human actions are determined by nature as a necessary stage in the purposeful evolution of nature towards perfection. I now briefly argue that the 'environmental crisis' of modernity is purely resultant from human actions.

Sloep and Dam-Mieras (2003, p.42) define an environmental problem as: "any change of state in the physical environment which is brought about by human interference with the physical environment, and has effects which society deems unacceptable in the light of its shared norms." This definition encapsulates a sliding scale of environmental problems from those that are local and temporary on the one hand, to those that are global and long-lasting on the other. The 'environmental crisis' as a concept has arisen because of the emergence in the last 100 years of an increasing number of environmental problems that are towards the global and long-lasting end of the scale. The 'environmental crisis' is thus purely resultant from the *human actions* which have created environmental problems that are characterised by their global reach and long-lasting nature.

This means that the above conclusion, that human actions are determined by nature as a necessary stage in the purposeful evolution of nature towards perfection, needs amending. As the 'environmental crisis' is purely resultant from human actions, it too must be part of this purposeful evolution. Therefore, the new conclusion that inevitably follows is: *the 'environmental crisis' is determined by nature as a necessary stage in the purposeful evolution of nature towards perfection.*

8.1.4 Objections to the reinterpretation

It could be objected that there are many references to human freedom in Hölderlin's work that would seem to cast doubt on the third claim. This is particularly noticeable in his novel Hyperion. For example, Hyperion (Holderlin, 1990, p.117) states that, "without freedom all is dead." However, this objection is easily answered because these references all appear in Hölderlin's early work, and even then they are more than counterbalanced by the opposing fatalistic views that I have outlined. In his early period Hölderlin was struggling to come to terms with the conflict between his keen moral aspirations for social change on the one hand, and his belief in perfection only arising through natural development on the other. In his later work, as is clear in his endorsement of the 'Hesperian' response to our condition, he firmly accepts the powers of natural development and the determination of human actions by nature. He realizes the futility of pursuing his idealistic moral aspirations because he accepts the illusory nature of human free will.

A further objection could be made that this reinterpretation is pointless because Darwin's theory of evolution, which emerged shortly after Hölderlin's time, gives a view of evolutionary processes that is incompatible with Hölderlin's view that there was a 'blessed unity of being' prior to the arrival of humans. We now know that the emergence of the human species – and its primordeal drive to art – was preceded by four billion years of evolution of life on Earth. It can thus be argued that there was not a 'blessed unity of being' prior to the evolution of humankind.

This is exemplified by the claim of Hans Jonas (2001, p.4) that the subject-object divide opened up four billion years ago, when, "living substance, by some original act of segregation, has taken itself out of the general integration of things in the physical context, set itself over against the world, and introduced the tension of "to be or not to be" into the neutral assuredness of existence." This certainly does not appear to be a pre-human 'blessed unity of being'. However, it is interesting that Jonas (2001, p.xv) also sees humans as, "a 'coming to itself' of original substance."

It is clear that this Darwinian based objection does not invalidate the views of Hölderlin, or the reinterpretation of them presented in this paper. In fact, not only does evolutionary theory perfectly complement Hölderlin's philosophy, his philosophy *needs* it. The idea that nature could use its power to instantaneously create a being as complex as a human out of the 'blessed unity of being' is hardly defensible. In the light of our knowledge today we can simply reinterpret Hölderlin as claiming that nature used its power four billion years ago to divide the 'blessed unity of being' and create a subject/object divide. As he sees nature as an unfolding and evolving organism, the divide would give rise to human subjects after a sufficient period of time. This, ""coming to itself" of original substance", as Jonas describes it, has in actuality taken approximately four billion years.

8.1.5 Conclusion

I have argued that the existing secondary literature has not grasped the full implications of Hölderlin's thought for what it means to be a human in modernity. By drawing together Hölderlin's ideas I have sought to understand his notion of the purpose of human actions, and what this purpose means for the 'environmental crisis'.

Hölderlin's conception of nature is an organism unfolding to perfection. I have argued that he sees modernity as an important stage of this unfolding, which is characterized by the development of technology through human actions. I have further argued that this means that the 'environmental crisis' of modernity – a side-effect of the development of technology – is also an inevitable stage of this unfolding; it is in the interests of nature. As nature continues to unfold, the disharmony of modernity will be succeeded by a re-conquered harmony. I have argued that Hölderlin's 'saving power' is actual technology, as this seems most consistent with his thought. Heidegger's view, that the 'saving power' is the essencing of technology, seems inconsistent with the positive role of technology in cosmic evolution that is envisioned by Hölderlin.

The reinterpretation I have outlined clearly entails an inversion of the conventional notion of causality in the 'environmental crisis' of modernity. Humanity is conventionally pictured as harming nature. My thesis has shown that for Hölderlin it is nature that is 'harming' humanity. We have been cast aside out of the rapture of nature into a realm of suffering and self-consciousness, with the purpose of developing technology to serve the purposes of the unfolding nature of which we are a part.

We are left with the question of what our attitudes to nature should be, given this reinterpretation of what it means to be a human in cosmic evolution. The answer is simple. As nature is the source of our individual attitudes, our attitudes to nature must be in the interests of nature. Our attitudes, whether they are techno-centric, environmentalist, quietist, or nature-exploitative are all correct for us as individuals, because in the aggregate they fulfil the purpose of nature as a whole.

Chapter 9

Conclusions

In the *Introduction* I claimed that we live in an epoch in which a violent clash exists. The interactions between humans and their surroundings typically lead to the belief that humans are radically different from their surroundings. However, human knowledge has increased to the point where humans have realised that they have evolved from their surroundings through a very gradual process of evolution. Given this knowledge one would not expect there to be radical differences between humans and their surroundings.

Throughout this book I have tried to reconcile this tension. It is possible that this attempt at reconciliation is misplaced. There are two reasons why this might be so. Firstly, there could be no surroundings; my first assumption in the *Introduction* was that this is false. Secondly, humans could actually be radically different from their surroundings; this could be true even if humans emerged from their surroundings through a very gradual process of evolution. However, such a 'radical difference' is much more plausible if humans did not so evolve. If the universe had a creator who created humans differently, or if humans evolved in an 'alternate universe' and then travelled to the Earth, then we would have a good reason to expect a great chasm to exist between humans and their surroundings.

Given these two possibilities one could just conclude that 'anything is possible' and stop thinking about the relationship between humans and their surroundings. However, if our *starting assumptions* are that human surroundings exist, and that humans evolved from these surroundings through a very gradual process of evolution, then our *starting expectation* should be that humans are extremely similar to their surroundings. This is my *starting expectation*. The question of interest thus becomes: Why do humans typically believe that they are radically different from their surroundings? My strategy throughout has been to address this question, to see if it rests on shaky foundations. If this belief has foundations which crumble when closely scrutinised then our tension can easily be reconciled. One can conclude that the widespread human belief that humans are radically different from their surroundings is misplaced and thereby resolve the tension in accordance with the *starting expectation*. On the contrary, if the foundations are strong then one might conclude that humans are radically different from their surroundings. My aim has been to expose the foundations and in so doing I have concluded that they are exceedingly crumbly.

In *Chapter Two* we concluded that humans cannot fully understand the nature of their surroundings because they are unable to access the interior states of their surroundings. In *Chapter Three* we saw that the conception of their surroundings that a human has is grounded in their perceptions of those surroundings. We can already see that the conceptions that a human has of their surroundings are likely to be inaccurate; for conceptions typically arise from perceptions and a human's perceptions are unable to access the interior states of their surroundings.

In *Chapter Three* we explored the various ways in which a human's perceptual apparatus is inevitably constrained. Firstly, it is unable to perceive the interior states of what is perceived.

Secondly, a perceptual apparatus has to perceive *in a particular way* – what is perceived is 'moulded', what is not perceived is 'filtered out'; so, much of what exists in human surroundings will not be perceived by humans. Thirdly, the human perceptual apparatus has in-built temporal constraints; it is only able to perceive movements from an exceptionally narrow temporal perspective. When one acknowledges these inevitable perceptual constraints then one can see that the widespread belief that humans are radically different from their surroundings rests on very shaky foundations.

This acknowledgement can cause one to start thinking about what one's surroundings are actually like, rather than how those surroundings appear to one to be. So, in *Chapter Three* I contended that the 'surroundings in-themselves' are a 'blobject' – a *continuous bringing forth of a different arrangement of particles and emptiness*. There are no objects in the 'blobject'; there are no objects in perceptions of the 'blobject'; objects come into existence for a particular part of the 'blobject' when that part conceptualises its perceptions in a certain way.[102] In

---

[102] Recall that I claimed that humans perceive *part of* the 'surroundings in-themselves' when they perceive the 'blobject'. The 'blobject' has a perceivable side and an unperceivable side ('feeling states', thoughts and awareness) and the perceivable side can be both perceived and conceptualised very differently. The perceptual divisions are dependent on the makeup of the perceptual apparatus of the perceiver, and the conceptual divisions are dependent on the thought processes that are occurring within the thinker. If one can get oneself into a particular state of thought (a state mostly lacking thought!) then one might be able to just perceive one's surroundings without conceiving of them as distinct objects. This would be easier if one was in very unfamiliar surroundings; humans spend most of their time in relatively familiar surroundings so they are always thinking the same thing: *'that is a cloud, that is a human, that is a tree, that is a book, that is an aeroplane, etc.'* The 'blobject'

Chapter Three we also considered some world views which entail that human surroundings are more similar to humans than initially appears to be the case. These views were 'panwhat-it-is-likeness' and 'panawareness' (which I took to include panexperientialism and panpsychism). In *Chapter Five* we encountered another view which entails that humans are similar to their surroundings – 'pansensism'.

In *Chapter Four, Chapter Five* and *Chapter Six* we considered various phenomena which are central to the relationship between humans and their surroundings. These phenomena exist in humans, the question we were concerned with is: Could these phenomena exist throughout human surroundings too? We found a common pattern; the contemporary conceptualisations of these phenomena are all 'thick' rather than 'thin'. What this means is that so much is 'packed in' to the concept that it seems implausible that the phenomena could exist throughout human surroundings. How did this come about? Why does this common pattern exist? This pattern arose precisely because human surroundings were assumed to be wholly devoid of these phenomena when the concepts were formed!

So, consciousness is widely hypothesised to be a singular state that contains both awareness and 'what-it-is-likeness'; this 'thick' view makes it seem unlikely that the phenomenon is pervasive in human surroundings. However, if one adopts a 'thin' conceptualisation then one can more plausibly believe that

---

isn't 'hidden from humans' it is just that it is very hard not to think of our surroundings as carved up, our normal state of existence is to think: *'what is that, what I can do with that, is that of any use to me, is that of any danger to me'*. Stop thinking and let the oneness of the 'blobject' reveal itself.

human surroundings are pervaded with 'what-it-is-likeness' but that they are not pervaded with awareness.

We saw that having 'a mind' is widely taken to be a very 'thick' phenomenon. It supposedly entails states such as 'feeling states', intentionality, freedom, causing certain movements, awareness, thinking and perception. If one takes all of these states to be part of having 'a mind' then one will naturally conclude that there is a chasm between humans and their surroundings; after all, who would believe that the part of human surroundings that is a stone thinks, perceives, is aware, has freedom, etc. However, if one adopts a 'thin' conceptualisation and asserts that having 'a mind' is just to think, then to believe that humans 'have a mind' but that a stone doesn't, clearly doesn't create such a great chasm. A stone could lack 'a mind' and still have some, or all, of the other attributes that many humans assume to be 'mental'.

We also saw that perception is still widely taken to be a 'thick' phenomenon which entails both the registering of information and the awareness of this. On the 'thin' conceptualisation one can believe that everything perceives (panperceptualism) without believing that everything is aware (panawareness).

Another interesting conclusion that we have reached is that scientific findings are often interpreted in order to make them compatible with the 'thick' conceptualisations. Indeed, the current obsession with the brain can be thought of as the most extreme 'thick' conceptualisation that one can have. On this view not only are human surroundings (non-brain human surroundings) wholly devoid of 'feeling states', colours, sounds, perceptions, intentionality, and awareness, but the non-brain human body is wholly devoid of these things too. All of these attributes are taken to be solely located in brains, and science is utilised to try and persuade people that if they disagree that they are wrong (*'you might swear that there is a stabbing pain in your left leg, but science proves you wrong!'*) – this brain obsession, this belief that the brain brings all of these attributes into existence, and that the non-brain universe is wholly devoid of these attributes, is the ultimate 'thick' conceptualisation.

Despite the dominance of the 'thick' conceptualisations we concluded that there are very good reasons to believe in the 'thin' conceptualisations – to believe that perception *can* exist without awareness, to believe that 'what-it-is-likeness' *can* exist without awareness, to believe that intentionality *can* exist without a mind, to believe that the feeling of pain *is* often located in one's non-brain body, etc.

In *Chapter Five* we considered how many senses a typical human has. I made the case for a reconceptualisation according to which a typical human only has two senses – seeing and hearing. According to this view what we refer to as the other three 'senses' are actually one phenomenon and could quite intelligibly pervade the entire universe. Furthermore, I made the case that the 'what-it-is-likeness' of the two human senses exists in human surroundings regardless of whether or not it is being perceived, and therefore that there is no distinction between humans and their surroundings arising from sensory 'what-it-is-likeness'.

In *Chapter Six* I contended that awareness is 'level-less' and has variable contents; this is in opposition to the wide-spread talk of 'altered states of consciousness' and different 'levels of consciousness'. I suggested that when one rids oneself of the idea that awareness can rise to different 'levels' or be 'altered' into different states then one will be less likely to believe that parts of human surroundings such as notebook computers are aware. I also contended that there is a close connection between awareness and sleep; this connection implies that the only parts of the universe that are aware are animals (most animals sleep). This conclusion was reinforced by suggesting that awareness arises due to 40-Hz oscillations which are generated in animal brains.

In *Chapter One* we considered the question of whether humans are natural and I contended both that humans are wholly natural and that they have unique attributes. However, I suggested that this latter conclusion is not particularly interesting because all parts of the universe have unique attributes. I also contended that the very fact that humans can doubt their naturalness is seemingly important. In *Chapter Seven* these themes were developed. I claimed that humans have a sense of 'specialness'; a sense of 'superiority' when contrasted to all of the other life-forms which live on the Earth; a sense of being 'not natural'; and that this is *because* humans are advanced tool-users. I also considered whether humans have a 'special' unique attribute which would in-itself (in the absence of the evolutionary perspective) justify talk of a great chasm between humans and the non-human universe, and I concluded that no such attribute exists.

In *Chapter Two* we considered the phenomenon of evolution and concluded that evolution is a fact and that we know quite a lot about evolution; however, our main conclusion was that the way that evolution occurs is fundamentally mysterious to us, largely because our perceptual apparatus is constrained in a way which means that the entire universe is fundamentally mysterious to us.

The question of whether the human species has a special relationship with the surroundings which evolved it was explored in *Chapter Eight*. I claimed that such a relationship exists, but that it can only exist if one takes the evolutionary perspective. This special relationship can exist if the human species evolved as a 'fluke'; however I contended that the universe is purposeful – that life is directed towards a 'human-like' species from the first moment that it evolves on a planet. I contended that given that the evolutionary process is fundamentally mysterious to us this claim is neither 'scientific' nor 'non-scientific'; however, it is highly plausible. We concluded in *Chapter Seven* that humans do not have 'special' unique attributes, so what does it mean to talk of a 'human-like' species? Why would life be directed towards, aiming for, a 'human-like' species? I suggested that what makes the human species 'special' stares every human in the face – it is technology, the same thing which gives rise to the sense of human specialness. To be 'human-like' is to be technological.

I claimed that a technological species is the pinnacle of the evolutionary progression of life on a planet. This is simply because life is a good state of feeling for the universe, and the universe strives/yearns to stay in a good state. This means that life needs to technologically leave the ageing Earth. In the short-term technology is required to maintain the temperature of the Earth's atmosphere because the non-technological homeostatic ability of the Earth is already weakening, and it has been further severely perturbed by the human mass movement of carbon from its underground storage areas to the thermohaline circulation.

There are certain requirements for becoming a technological species. Before becoming technological a species needs to be an advanced-tool using species. In order to develop advanced tools a species needs to be able to think about its surroundings and explore them to see how it can most fruitfully modify them. In order to make the transition from advanced tool-using to technological a species needs to take this thinking and exploration to a new level – it needs to become scientific. A prerequisite for science and technology is also a particular mindset – a mindset of 'specialness' of 'non-naturalness': *our surroundings are very different from us – they are 'nature', they are 'matter', they can be understood by us and moulded by us into new forms.* I claimed in the *Introduction* that: the human philosophical endeavour is itself a stage in the evolution of the universe. We can now see that it is a very important stage – humans needed to become philosophical in order to become scientific and technological.

> The human: the doubter of its naturalness; the advanced-tool user which evolved into the bringer forth of technology; the philosophiser; the saviour of life on Earth.

Let us return to the three interrelated aspects of our exploration. Firstly, we have been considering why contemporary humans typically consider themselves to be radically different from their surroundings. We have seen that the answer to this question has two elements. Firstly, the human perceptual apparatus operates in such a way as to make this outcome likely. Secondly, humans are that part of life which is technological, and being technological gives rise to a sense of 'specialness', of 'separateness'. As Hölderlin puts it, the epoch of human technological development can be characterised as '*the absence of the gods'*. It is this 'technological separateness' which explains the dominance of the 'thick conceptualisations' and the contemporary obsession with the brain; in other words, it explains why contemporary humans typically consider themselves to be radically different from their surroundings.

Secondly, we have been considering the likelihood that humans could actually be very similar to their surroundings. We have concluded that there are very good reasons to believe that humans could be much more similar to their surroundings than most humans ordinarily assume. I have defended the idea that the 'feeling states' that a human becomes aware of are located throughout their body (not just in their brain), and from the evolutionary perspective there seems to be no good reason to deny that these states could pervade the entire universe (which is the panwhat-it-is-likeness view). I have claimed that panwhat-it-is-likeness is a more appealing position that panawareness. This is because panawareness advocates conclude that there are different levels of 'what-it-is-likeness'=awareness and thereby face the 'combination problem'. In contrast, panawareness doesn't involve levels – there are simply level-less 'what-it-is-likeness' states and level-less awareness states.

I have contended that every 'feeling state' is a perception, so this entails that panperceptualism is true as well as panwhat-it-is-likeness. Furthermore, a perception is a perception *of* something, so this also entails that the entire universe is pervaded with intentionality. I have also contended that the 'what-it-is-likeness' of the two human senses is located in human surroundings, this means that humans are not divided from their surroundings through the phenomenon of 'what-it-is-likeness (neither 'feeling states' nor 'sensory what-it-is-likeness'). I have suggested that arrangements of the universe such as 'video cameras' and 'mobile phones' are 'senses'; this means that many parts of the living and non-living universe have 'senses'. Finally, I have defended the idea that humans are wholly natural and so are not divided from their surroundings by the phenomenon of naturalness.

However, I have also suggested that there are potentially significant differences between humans and their surroundings. I have supported the idea that awareness is a discontinuous phenomenon which pops into and out of existence, and that only particular arrangements of the universe have the capacity to generate this popping. This implies that there is a difference of great significance between humans and the *vast majority* of their surroundings. This potentially warrants talk of there being a 'chasm' between humans and most of their surroundings, although it is a far smaller chasm than if one has a 'thick' conceptualisation of consciousness. And, of course, this chasm is not between humans and their surroundings; it is a chasm which has humans and the overwhelming majority of non-human animals on one side, and the rest of the universe on the other side. Furthermore, the human capacity for advanced thought seems to be a potentially significant difference between humans and the vast majority of their surroundings; although as with awareness, many non-human animals seem to be on the human side of the division.[103]

---

[103] It is possible that one might still be trying to rationalise one's *sense of human specialness* through believing that a human is superior to all of the non-human animals that live on the Earth solely because a human has 'advanced thought'. However much evidence accrues that non-human animals have advanced thought and language, there are still likely to be humans who assert *'humans are special because human thought/language is much more advanced'*. The sense of

Let us consider the 'blobject'. It is pervaded with 'what-it-is-likeness' – 'feeling states', colours and sounds – and every interaction, every movement, involves a change in 'what-it-is-likeness'. The 'blobject' is also pervaded with states of intentionality and states of perception. In certain, relatively rare, parts of the 'blobject' processes of thought are occurring (in these parts 'a mind' exists); and when these thoughts conceptualise perceptions the 'blobject' is carved up into objects. In some relatively rare parts of the 'blobject' awareness momentarily pops into existence.

This nicely takes us to the question of what it means to be a human. Recall that in *Chapter One* I stated:

> Perhaps *to be human* is something that transcends biological classification. Perhaps *to be human* is to be a part of nature that considers itself to be not natural. If this is so, then when a particular biological entity loses its sense of opposition – of separateness – of non-naturalness – then it will cease to be human.

From the perspective of the 'blobject' we can now make more sense of this. There are various areas of thought – processes of thought – in the 'blobject'. Wherever there is an area of thought that considers itself to be not natural, then this area of thought is a human. This 'consideration' *requires* an 'area of thought' but it is *not itself* a thought; it arises from

---

human specialness is a very hard thing to shake off (and with good reason, because given that we live in an evolving universe a technological species *is* special!).

accumulating a broad range of knowledge about its surroundings, but itself it is simply the 'sense of human specialness'
which we explored in *Section 7.1*.[104] There is another sense of
human – the 'biological human'; this term refers not to the
'surroundings in-themselves' but to the perceivable part of the
universe which is conceptualised by humans when they think
about their perceptions; the term refers to those parts of their
surroundings that a human conceptualises as 'human' – biological entities which typically have legs, arms, a torso and a head.

One should keep in mind that humans do not know the
nature of their surroundings. So, any conclusions that are made
regarding how similar 'biological humans' are to their surroundings are conclusions which are based on what seems to be the
most plausible thing to believe. The most plausible thing to
believe seems to me to be that the entire universe is fundamentally similar. However, animals seem to have two particularly
interesting attributes – level-less awareness and thought (in
varying degrees of advancement). It seems obvious to me that a
multitude of species of animals have both awareness and
advanced thought; and it is plausible to believe that humans
have more advanced thought than any other species of animal
on the Earth. However, given our limitations we should accept
the possibility that thought actually pervades the universe. And,
despite the sharp distinction that I have sought to make between
perception and awareness, one has to accept the possibility that
a panperceptualism universe is actually a panawareness universe. I have suggested that humans are distinguished from the
majority of their surroundings through just two attributes –

---

[104] Human = advanced tool user = 'sense of human specialness' = considers itself
to be not natural.

thought and awareness – however, humans could be even more similar to their surroundings than I have suggested.

This brings us nicely to the third aspect of our exploration which is to consider whether or not the human species has a special place in the evolutionary process. It might be true that humans and non-human animals have seemingly interesting attributes – awareness and thought – which the rest of the universe lacks, but is there anything uniquely special about *just* the human species? I have claimed that unless one takes the evolutionary perspective it is hard to defend the idea that the human species is special. However, I have claimed that humans do have a uniquely special place in the evolutionary progression of life because they are a technological species. And becoming technological seems to require being aware, a high level of thought, and a host of other human attributes. This means that we can make rational sense of the idea that the attributes which seemingly create a chasm between humans and *most* of their surroundings – awareness and advanced thought – are special attributes. We can also understand why this chasm exists; it is a necessary prerequisite for the development of technology.

We have exposed the shaky foundations which underpin the idea that humans are radically different from their surroundings, and we have explored numerous views which entail varying degrees of increased similarity between humans and their surroundings. We can conclude that the most plausible thing to believe is that humans are much more similar to their surroundings than most humans ordinarily assume. To accept this is not to believe that humans are not special. For, the evolutionary perspective which underpins this conclusion also leads to another conclusion: the human species has a very special place in the evolution of life. Indeed, the human species is the pinnacle of the evolutionary process on the Earth – a technological species which is necessarily fundamentally opposed to that to which it is so similar.[105]

---

[105] The pinnacle of the evolutionary process on a planet is a technological species. Technology creation is the pinnacle. The pinnacle is not intelligence. There could currently be more intelligent life-forms on the Earth than humans, or in the future humans could technologically create entities which are more intelligent than humans. None of this changes the fact that the human species, as the part of life on Earth that has become technological, is the pinnacle of the evolutionary process on the Earth.

Bibliography

Arendt, H., 1978. *The Life of the Mind – Two/Willing*. London: Secker and Warburg.

Aristotle, 1980. On Colours. In *Aristotle: Minor Works*. Translated by W. S. Hett. London: William Heinemann Ltd.

Aristotle, 2001. De Anima. In R. McKeon, ed. *The Basic Works of Aristotle*. Bk. III Ch. 1. New York: The Modern Library.

Armstrong, D.M., 1963. Vesey on Sensations of Heat. *Australasian Journal of Philosophy*, 41, p.361.

Armstrong, D.M., 1968. *A Materialist Theory of the Mind*. London: Routledge.

Armstrong, D.M., 1980. *The Nature of Mind and Other Essays*. St Lucia: University of Queensland Press.

Armstrong, D.M., 1981. *The Nature of Mind*. USA: Cornell University Press.

Armstrong, D. M., 1997. What is Consciousness? In Block, N., Flanagan, O. and Guzeldere, G. eds. 1997. *The Nature of Consciousness*. London: MIT Press.

Baars, B.J., 1997. *In The Theater of Consciousness*. Oxford: Oxford University Press.

Ball, P., 2008. *Bright Earth: The Invention of Colour*. London: Vintage.

Barlow, C., 1991. *From Gaia to Selfish Genes: Selected Writings in the Life Sciences*. London: MIT Press.

Baron-Cohen, S., 1996. *Mindblindness: An Essay on Autism and Theory of Mind*. London: MIT Press.

Block, N., 1997. On a Confusion about a Function of Consciousness. In Block, N., Flanagan, O. and Guzeldere, G. eds. 1997. *The Nature of Consciousness*. London: MIT Press.

Block, N., 2004. Consciousness. In Guttenplan, S. ed. 2004. *A Companion to the Philosophy of Mind*. Bodmin: Blackwell.

Bohm, D., 1990. A new theory of the relationship of mind and matter. *Journal of Philosophical Psychology*, 3(2), p.283.

Bornstein, R.F., 1992. Perception Without Awareness: Retrospect and Prospect. In Bornstein R.F. and Pittman T.S. eds. 1992. *Perception Without Awareness*. London: Guilford Press.

Bucke, R.M., 1905. *Cosmic Consciousness: A Study in the Evolution of the Human Mind*. Philadelphia: Innes & Sons.

Campbell, J., 2001. A Simple View of Colour. In Byrne, A. and Hilbert, D.R. eds. *Readings on Color: The Philosophy of Color*. Volume 1. London: MIT Press.

Chalmers, D., 1996. *The Conscious Mind*. Oxford: Oxford University Press.

Coburn, R., 1966. Pains and Space. *Journal of Philosophy*, 63(13), p.395.

Conway Morris, S., 2005. *Life's Solution: Inevitable Humans in a Lonely Universe*. Cambridge: Cambridge University Press.

Crick, F. and Koch, C., 1997. Towards a Neurobiological Theory of Consciousness. In Block, N., Flanagan, O. and Guzeldere, G. eds. 1997. *The Nature of Consciousness*. London: MIT Press.

Dawkins, R., 1999a. *The Selfish Gene*. Oxford: Oxford University Press.

Dawkins, R., 1999b. *The Extended Phenotype*. Oxford: Oxford University Press.

Dawkins, R., 2006. *The Blind Watchmaker*. London: Penguin Books.

Dennett, D., 1991. *Consciousness Explained*. Boston: Little Brown.

Descartes, 1996. In J. Cottingham, ed., *Descartes: Meditations on First Philosophy*. Cambridge: Cambridge University Press.

Dickens, C., 1969. *Hard Times*. London: Penguin Books.

Dixon, N.F., 1971. *Subliminal Perception - The Nature of a Controversy*. London: McGraw-Hill.

Feenberg, A., 2003. Critical Evaluation of Heidegger and Borgmann. In Scharff, R.C. and Dusek V. eds. 2003. *Philosophy of Technology: The Technological Condition – An Anthology*. Oxford: Blackwell Publishing.

Feigl, H., 1967. *The "Mental" and the "Physical"*. London: Oxford University Press.

Gatens, M., 1996. *Imaginary Bodies: Ethics, Power and Corporeality*. London: Routledge.

Godfrey-Smith, P., 2001. On the Status and Explanatory Structure of Developmental Systems Theory. In Oyama, S., Griffiths P.E. and Gray, R.D. eds. 2001. *Cycles of Contingency*. London: The MIT Press.

Goff, P., 2006a. Experiences Don't Sum. In Strawson, G., 2006. *Consciousness and its place in nature*. Exeter: Imprint Academic.

Goff, P., 2006b. *Should Materialists be afraid of Ghosts*. Ph. D. University of Reading.

Goldie, P., 2002. *The Emotions: A Philosophical Exploration*. Oxford: Oxford University Press.

Goodale, M. and Milner, D., 2004. *Sight Unseen*. Oxford: Oxford University Press.

Greene, B., 2000. *The Elegant Universe*. London: Vintage.

Gregory, R.L., 1987. ed. *The Oxford Companion to the Mind*. Oxford: Oxford University Press.

Gregory, R.L., 2004. ed. *The Oxford Companion to the Mind*. 2nd Edition. Oxford: Oxford University Press.

Griffin, D.R., 1998. *Unsnarling the World-Knot*. London: University of California Press.

Griffin, D.R., 2007. *Unsnarling the World-Knot*. Eugene: Wipf and Stock.

Griffiths, P.E. and Gray, R.D., 2001. Darwinism and Developmental Systems. In Oyama, S., Griffiths P.E. and Gray, R.D. eds. 2001. *Cycles of Contingency*. London: The MIT Press.

Hardin, C.L., 1988. *Color for Philosophers*. Cambridge: Hackett Publishing Company.

Haverkamp, A., 1996. *Leaves of Mourning: Hölderlin's Late Work*. New York: SUNY Press.

Heidegger, M., 1956. *Existence and Being*. London: Vision Press Ltd.

Heidegger, M., 2003. The Question Concerning Technology. In Scharff, R.C. and Dusek, V. eds. 2003. *Philosophy of Technology: The Technological Condition – An Anthology*. Oxford: Blackwell Publishing.

Heil, J., 1998. *Philosophy of Mind: A Contemporary Introduction*. London: Routledge.

Hölderlin, F., 1988a. The Perspective from which We Have to Look at Antiquity. In Pfau, T. ed. 1988. *Friedrich Hölderlin: Essays and Letters on Theory*. New York: SUNY Press.

Hölderlin, F., 1988b. No.41: To his Mother. In Pfau, T. ed. 1988. *Friedrich Hölderlin: Essays and Letters on Theory*. New York: SUNY Press.

Hölderlin, F., 1990. Hyperion. In Santner E.L. ed. 1990. *Hyperion and Selected Poems*. New York: Continuum.

Hölderlin, F., 2003. Being Judgement Possibility. In Bernstein J.M. ed. 2003. *Classic and Romantic German Aesthetics*. Cambridge: Cambridge University Press.

Hudson, H., 1961. Why are our feelings of pain perceptually unobservable? *Analysis*, 21.5, p.99.

Hull, D.L., 1998. Introduction to Part IV: Species. In Hull, D.L. and Ruse, M. 1998. *The Philosophy of Biology*. Oxford: Oxford University Press.

Hyman, J., 2003. Pains and Places. *Philosophy*, 78.

Jablonka, E., 2001. The Systems of Inheritance. In Oyama, S., Griffiths P.E. and Gray, R.D. eds. 2001. *Cycles of Contingency*. London: The MIT Press.

Jablonka, E. and Lamb, M.J., 2005. *Evolution in Four Dimensions*. London: The MIT Press.

James, W., 1912. *Essays in Radical Empiricism*. London: Longmans, Green, and Co.

James, W., 1950. *The Principles of Psychology - Volume 1*. New York: Dover Publications Inc.

Jonas, H., 2001. *The Phenomenon of Life: Toward a Philosophical Biology*. Illinois: Northwestern University Press.

Kant, I., 1992. Dreams of a Spirit-Seer. In Walford, D. ed. 1992. *Theoretical Philosophy 1755-1770*. New York: Cambridge University Press.

Koch, C., 2005. In Blackmore, S., 2005. *Conversations on Consciousness*. Oxford: Oxford University Press.

Korsmeyer, C., 2009. Disputing Taste. *The Philosophers Magazine*, Issue 45, 2nd Quarter, p.71.

LaBerge, S., 2005. In Blackmore, S., 2005. *Conversations on Consciousness*. Oxford: Oxford University Press.

Laszlo, E., 2004. *Science and the Akashic Field*. Vermont: Inner Traditions.

Lewontin, R., 2000. *The Triple Helix*. London: Harvard University Press.

Llinas, R., 2002. *i of the vortex – From Neurons to Self*. Massachusetts: The MIT Press.

Locke, J., 1975. *An Essay Concerning Human Understanding*. P. H. Nidditch ed., (1690/1975) %9-10. Oxford: Oxford University Press.

Lorenz, K., 1977. *Behind the Mirror*. Translated by R. Taylor. Suffolk: Methuen & Co. Ltd.

Lorenz, K., 1982. Kant's doctrine of the a priori in the light of contemporary biology. In Plotkin, H. C. ed. 1982. *Learning, Development and Culture: Essays in Evolutionary Epistemology*. Chichester: John Wiley & Sons.

Lovelock, J., 2000. *Gaia: A new look at life on Earth*. Oxford: Oxford University Press.

Lovelock, J., 2006. *The Revenge of Gaia*. London: Penguin Books Ltd.

Lowe, E.J., 2000. *An introduction to the philosophy of mind*. Cambridge: Cambridge University Press.

Lycan, W.G., 1996. *Consciousness and Experience*. London: MIT Press.

Maclachlan, D.C.L., 1989. *Philosophy of Perception*. Englewood Cliffs: Prentice Hall.

Margulis, L., 2001. *The Symbiotic Planet: A new look at evolution*. Guernsey: Phoenix.

Margulis, L. and Sagan, D., 1997. *Microcosmos: Four billion years of microbial evolution*. London: University of California Press.

Margulis, L. and Sagan, D., 2002. *Acquiring Genomes: A theory of the origins of species*. New York: Basic Books.

Matthews, E., 2005. *Mind*. London: Continuum.

McGinn, C., 1982. *The Character of Mind*. Oxford: Oxford University Press.

McGinn, C., 1989. Can we solve the mind-body problem? *Mind*, 98, pp.349-50.

McGinn, C., 1999. *The Mysterious Flame*. New York: Basic Books.

Molnar, G., 2003. *Powers: A Study in Metaphysics*. Oxford: Oxford University Press.

Murdoch, I., 1973. *The Black Prince*. London: Penguin Books.

Nauen, F. G., 2001. *Revolution, Idealism and Human Freedom: Schelling, Hölderlin and Hegel and the Crisis of Early German Idealism*. USA: Indiana University Press.

Newton, I., 1952. *Opticks, Or a Treatise of the Reflections, Refractions, Inflections and Colours of Light*. 1730/1952. New York: Dover Publications.

Ngoc, T. 2010. Thanhniennews [online] Available at: http://thanhniennews.com/features/?catid=10&newsid=1267 3 [accessed 10 February 2010].

Noe, A., 2009. *Out of Our Heads*. New York: Hill and Wong.

Nudds, M., 2009. Discriminating Senses. *The Philosophers Magazine*, Issue 45, 2nd Quarter, p.92.

O'Callaghan, C., 2009. The world of sounds. *The Philosophers Magazine*, Issue 45, 2nd Quarter, pp.64-65.

Oxford English Dictionary, 2009. [online] Available at: http://dictionary.oed.com/cgi/entry/50175079?single=1&query_type=word&queryword=perceive&first=1&max_to_show =10 [Accessed 8 February 2009]

Oyama, S., 2001. Terms in Tension: What Do You Do When All the Good Words Are Taken. In Oyama, S., Griffiths, P.E and Gray, R.D. eds. 2001. *Cycles of Contingency*. London: The MIT Press.

Oyama, S., Griffiths, P.E. and Gray, R.D. eds., 2001. *Cycles of Contingency*. London: The MIT Press.

Papineau, D., 2004. *Thinking about Consciousness*. Oxford: Oxford University Press.

Papineau, D., 2005. *Introducing Consciousness*. Cambridge: Icon Books, 2005.

Peacock, R., 1938. *Hölderlin*. London: Methuen & Co. Ltd.

Peake, S., 2003. A citizen's guide to climate science. In Peake, S., and Smith, J. eds. 2003. *Climate Change: From Science to Sustainability*. Milton Keynes: The Open University.

Pfau, T., 1988. *Friedrich Hölderlin: Essays and Letters on Theory*. New York: SUNY Press.

Pierce, J.R. and David, E.E., 1958. *Man's World of Sound*. New York: Doubleday & Company Inc.

Priest, S., 1991. *Theories of the Mind*. London: Penguin Books.

Robinson, W.S., 1988. *Brains and People: An Essay on Mentality and Its Causal Conditions*. Philadelphia: Temple University Press.

Rosenberg, G., 2004. *A Place for Consciousness*. Oxford: Oxford University Press.

Rosenthal, D., 1997. A Theory of Consciousness. In Block, N., Flanagan, O. and Guzeldere, G. eds. 1997. *The Nature of Consciousness*. London: MIT Press.

Rosenthal, D., 2005. *Consciousness and Mind*. Oxford: Oxford University Press.

Ruse, M., ed. 1989. *Philosophy of Biology*. New York: Macmillan Publishing Company.

Russell, B., 1956. *Portraits from Memory*. Spokesman.

Scarry, E., 1985. *The Body in Pain*. Oxford: Oxford University Press.

Scharff, R.C. and Dusek, V., 2003. Introduction to Heidegger on Technology. In Scharff R.C. and Dusek V. eds. 2003. *Philosophy of Technology: The Technological Condition – An Anthology*. Oxford: Blackwell Publishing.

Schmidt, D.J., 2001. *On Germans and Other Greeks*. USA: Indiana University Press.

Seager, W., 1995. Consciousness, Information and Panpsychism. *Journal of Consciousness Studies*, 2(3), p.280.

Seager, W., 1999. *Theories of Consciousness*. London: Routledge.

Searle, J. R., 1992. *The Rediscovery of the Mind*. Cambridge: MIT Press.

Searle, J.R., 1997. *The Mystery of Consciousness*. London: Granta Books.

Searle, J.R., 2002. *Consciousness and Language*. Cambridge: Cambridge University Press.

Searle, J.R., 2005. In Blackmore, S. ed. 2005. *Conversations on Consciousness*. Oxford: Oxford University Press.

Sjolander, S., 1997. On the Evolution of Reality – Some Biological Prerequisites and Evolutionary Stages. *Journal of Theoretical Biology*, 187.

Skrbina, D., 2005. *Panpsychism in the West*. London: The MIT Press.

Sloep, P. B. and Dam-Mieras, M.C.E., 2003. Science on Environmental Problems. In Glasbergen, P. and Blowers, A. eds. 2003. *Environmental Policy in an International Context: Perspectives*. Oxford: Butterworth-Heinmann.

Spinoza, B., 1966. LVIII: To Schuller. Trans. Wolf, A. ed. 1966. *The Correspondence of Spinoza*. 2nd ed. London: Frank Cass & Co. Ltd.

Stewart, R., 2009. *Introduction to Physical Oceanography*. Florida: Orange Grove Texts Plus.

Stoljar, D., 2009. Physicalism, *The Stanford Encyclopedia of Philosophy* [online] Available at: <http://plato.stanford.edu/archives/fall2009/entries/physic alism/>   [Accessed 25 May 2010].

Stone, A., 2003. Irigaray and Hölderlin on the Relation Between Nature and Culture. *Continental Philosophy Review*, 36(4).

Stone, A., 2005. *Nature in Continental Philosophy – Week 4, Section V, Friedrich Hölderlin* [online] Available at: <http://www.lancaster.ac.uk/depts/philosophy/awaymave/408 new/wk4.htm> [Accessed 25 October 2005].

Strawson, G., 1994. *Mental Reality*. London: MIT Press.

Strawson, G., 2006. *Consciousness and its place in nature*. Exeter: Imprint Academic.

Strawson, G., 2008. *Real Materialism and other essays*. Oxford: Oxford University Press.

Strawson, G., 2009. Realistic Monism. In Skrbina, D. ed. 2009. *Mind that Abides*. Philadelphia: John Benjamins Publishing Co.

Swinburne, R., 1988. *The Evolution of the Soul.* Oxford: Clarendon Press.

Tarnas, R., 2007. *Cosmos and Psyche.* USA: Plume.

Tart, C.T., 1990. *Altered States of Consciousness.* 3rd Edition. New York: Harper Collins.

Taylor, C., 1994. *Sources of the Self: The Making of the Modern Identity.* Massachusetts: Harvard University Press.

Thompson, E., 1995. *Colour Vision.* London: Routledge.

Tye, M., 1997. A Representational Theory of Pains and Their Phenomenal Character. In Block, N., Flanagan, O. and Guzeldere, G. eds. 1997. *The Nature of Consciousness.* London: The MIT Press.

Tye, M., 2007. The Problem of Common Sensibles. *Erkenntnis*, 66, p.293.

Unger, R., 1984. *Friedrich Hölderlin.* Boston: Twayne Publishers.

Unger, R., 2006. *All the Power in the World.* Oxford: Oxford University Press.

Velmans, M., 2007. Dualism, Reductionsim, and Reflexive Monism. In Velmans, M. and Schneider, S. eds. 2007. *The Blackwell Companion to Consciousness.* Oxford: Blackwell.

Vollmer, G., 1984. Mesocosm and Objective Knowledge. In Wuketits, F. M. ed. 1984. *Concepts and Approaches in Evolutionary Epistemology*. Dordrecht: D. Riedel Publishing Co.

Weber, B.H. and Depew, D.J., 2001. Developmental Systems, Darwinian Evolution, and the Unity of Science. In Oyama, S., Griffiths, P.E. and Gray, R.D. eds. 2001. *Cycles of Contingency*. London: The MIT Press.

Wilczek, F. and Devine, B., 1988. *Longing for the Harmonies: Themes and Variations from Modern Physics*. London: W.M. Norton & Company.

Wilson, C., 2006. Commentary on Galen Strawson. In Strawson, G., 2006. *Consciousness and its place in nature*. Exeter: Imprint Academic.

Wittgenstein, L., 1989. *Philosophical Investigations*. Trans. G. E. M. Anscombe. Oxford: Blackwell.

Wittgenstein, L., 1968. Notes for Lectures on "Private Experience" and "Sense Data". *The Philosophical Review*, 77(3), p.318.

Wittgenstein, L. 1980. *Remarks on The Philosophy of Psychology. Vol. II*, von Wright, G.H. and Nyman, H. eds. 1980. Trans. C.G. Luckhart and M.A.E. Aue. Oxford: Blackwell.

Wittgenstein, L. 2006. "The Inner and the Outer". In Kenny, A., 2006. *The Wittgenstein Reader*. 2ND Edition. Oxford: Blackwell.

Weiskrantz, L., 1992. Introduction: Dissociated Issues. In Milner, A. D. and Rugg M.D. eds. 1992. *The Neuropsychology of Consciousness*. London: Academic Press Ltd.

Weiskrantz, L., 1999. *Consciousness Lost and Found*. New York: Oxford University Press.

Wyller, T., 2005. The Place of Pain in Life. *Philosophy*, 80, p.385.